"Musicians shouldn't be intimidated by the [...] *The Physics and Neuroscience of Music*. This is [...] music fan will find both enjoyable and educational. The que[...] ing the science, biology, and math related to music are made easily understandable, and the book is grounded in David's passion for both creating and enjoying music. At the end, anyone reading this book will have a greater appreciation for the creative spirit and a way to understand music in even deeper ways."

— Bob Neuwirth, singer-songwriter and record producer

"Putting the worlds of science and music together is an ambitious and potentially intimidating endeavor. But David Sulzer had me at paragraph one, where he writes 'no one needs this book'! No, I don't need it—but I find I do want it."

— John Schaefer, host of *New Sounds*, WNYC

"When your band protests, 'Whaddaya mean "dynamics"? I'm playing as loud as I can!'—turn them onto the solid matter in *Music, Math, and Mind*. As to Soldier's confection? A ribald reality check on what makes music matter and why we should mind. I've waited seventy-six years in a musical immersion to put a buzz on Dave Soldier's flyleaf."

— Van Dyke Parks, performer, arranger, producer, composer, and lyricist, including with the Beach Boys

Nota Bene

This book is for musicians and art lovers who may have had little exposure to math, physics, and biology. It is written so that you can understand everything with no math beyond grade-school multiplication and division.

This is not a "pop science" book to be absorbed in a single reading: read a chapter that you are interested in, absorb what you can, and when you reread it, you will understand more. Honestly, I have been a professional scientist for three decades and still need to learn and relearn many of the basic concepts in these pages.

To some extent, math and science are foreign languages, and it is best to learn little by little: some topics required humanity centuries to comprehend, and particularly toward the end of the book, there remains far more to be discovered.

For ambitious readers, the "Math Boxes" use simple math to go a bit further and can be safely ignored by readers without losing the flow. Even these require only multiplication and division.

The "Sidebars" are tangential remarks.

MUSIC, MATH, AND MIND

• • • •

Introduction

• Who needs a book on math's and the nervous system's roles in music?

Some of the questions that will introduce these chapters lack a clear answer, but this has one: *no one needs this book.*

The creation and appreciation of music do not require knowledge of the math and biology that allow it to exist. These topics are not taught to music students, and musicians create great work without knowing them. Still, understanding the basis of music, sound, and perception will explain some mysteries, open deeper ones, help you understand what you hear, and provide ideas for your own work.

The approach to contemporary musical education is rooted in a system developed to train orphans and abandoned children in Renaissance Naples. The word *conservatori* meant "places to save children," and music provided a way for those who did not inherit a family trade to learn to compose, play instruments, and sing to make a living. The original conservatory, Santa Maria di Loreto, founded in Naples in 1537, was immensely successful, training the composers Alessandro Scarlatti and Domenico Cimarosa. The movement spread, with Antonio Vivaldi teaching at the Ospedale della Pietà for orphaned and abandoned girls in Venice and composing concerti like the *Four Seasons* for the school orchestra. The system migrated to the Paris Conservatory in 1784 and from there throughout the world as conservatories broadened their mission to accept students from any family background.

Over these five hundred years, the purpose of musical training has always been to impart tools to make a living. For example, Johann Sebastian Bach, a god for composers, taught music theory classes at the Leipzig Thomas School so that students could perform on the organ during church services. Bach's lessons were bound in a book, *Precepts and Principles for Playing the Through Bass or Accompanying in Four Parts* (1738), modeled on the *Musical Guide* (1710) by Friedrich Erhard Niedt, a pupil of Bach's cousin Johann Nicolaus Bach. These books provide clear instructions in the techniques used by the Bach family, with rules for composing fugues and the popular dances like sarabandes and jigs: learning those rules can help you create a piece in that tradition.

In contrast, the creation of music in some other styles requires no theory classes—or even instrument or singing lessons—but simply talent, opportunity, and work. A pioneer of this approach was the French radio engineer Pierre Schaeffer, who in the 1940s composed exciting music by splicing together recording tape. Fifty years later, this approach was updated by the hip-hop group Public Enemy, who created instrumental tracks entirely by juxtaposing previously recorded sounds. Schaeffer and Public Enemy prove that with perseverance and access to the right tools, one can create brilliant music immediately without requiring years of training.

A thoughtful perspective on the issue of technical training versus instinct in creating music comes from the American composer and violinist Leroy Jenkins. Leroy grew up in the 1940s performing classical violin duos on the Chicago streets with his classmate, the future rock 'n' roll pioneer Bo Diddley, when lessons on instruments were typically part of the public school curriculum. Many of Leroy and Bo's classmates at Chicago's DuSable High School trained under the violinist and teacher Walter Dyett. Dyett's students included a high percentage of the top figures in jazz and pop, including the saxophonists Eddie Harris, Gene Ammons, and Clifford Jordan and the singers Dinah Washington, Johnny Hartman, and Nat King Cole.

Despite the success of Walter Dyett and others, starting in the 1970s, support for music lessons in many American public schools evaporated. Young people, like the Bronx DJs Kool Herc and Afrika Bambaataa, adapted the sparse instrumental resources available to them—drum machines and turntables from stereo supply stores—and used them to create a style that Bambaataa named hip-hop, presently the most popular musical style in the world.

Leroy's comment was that they, like his generation, developed a style from what was available because even in a desert of resources, "you can't kill human creativity."

An extreme example of creating music without training or theory would be music made by children the first time they are given musical instruments. In the 1960s, the electronic music pioneer Daphne Oram did just this by coaching British high school students to compose with kitchen and household appliances like vacuum cleaners—and they made a wonderful record.

Another approach is to ask children what music they like to listen to and then coach them on how to do something like it. With friends and collaborators, I have managed this with groups of children as young as three-year-olds, in Brooklyn (the Tangerine Awkestra), East Harlem (Da Hiphop Raskalz), and the Mayan highlands of Guatemala (Yok K'u). Each group made its own record in styles they loved, while playing instruments they handled for the first time. Their music is unconcerned with virtuosity and very aware of feelings and stories, and it can be as satisfying to listen to as work by professionals, who wouldn't be able to unlearn their craft to create the same way.

Whether music is made after years of training and knowledge or by children playing instruments for the first time, underlying all of this activity is an ancient and direct line of study initiated by natural philosophers in China, India, Egypt, and Greece. Their discoveries underpin the creation and comprehension of music but are unknown to almost all musicians.

Here are some basic questions on math, physics, and the nervous system that musicians care about but are not discussed in music theory classes:

- Which sounds are in and out of tune, and how are musical scales derived?
- Is it true that scales are really never in tune?
- What are overtones and harmonic sounds?
- Sound is formed from air waves that move in space and time. What shape are these sounds? How big, fast, and heavy are they? How are sound waves different in air, under water, or when traveling through the earth?
- Why do voices and instruments sound different from each other? Why do larger instruments play lower pitches?
- Given that we have only two eardrums and two ears, how are we able to identify many sounds that occur simultaneously?
- Is there a mathematical definition of noise and of consonance?
- How does the brain understand what it is listening to? (Warning: this does not yet have a satisfying answer.)
- How are emotions carried by music? (The same warning applies.)
- How do other animals hear and make sound differently from us?

If these issues are not taught to musicians and music lovers, it is not from a lack of curiosity. Artists and art lovers have of plenty of that, and this book is for them. You can't kill human creativity, and you might use this knowledge for new explorations.

Introductory Listening

Pierre Schaeffer's "Etude aux casseroles, dite 'pathetique'" was composed on recording tape in 1948. He created this music by assembling prerecorded sounds, including bits of a harmonica played by the blues musician Sonny Terry. It still sounds like music from the future.

Daphne Oram was also a pioneer of electronic music. For a project in 1968, likely inspired by John Cage, who used transistor radios as musical instruments, she coached English schoolchildren to create music using household appliances. Listen to "Adwick High School Number 3" by Linda Parker, a piece for metal sheet, tin lips, chimes, wire, frame, keyboards, wood block, echo sander, scraper, and whistle.

It's not hard to find "instrumentals" of vintage hip-hop tracks, that is, the musical portions without vocals. The first hip-hop hit, 1979's "Rapper's Delight" by the Sugar Hill Gang, was produced by the rock 'n' roll and disco singer Sylvia Robinson (her group Mickey and Sylvia had a major hit in the 1950s, Bo Diddley's song "Love Is Strange") and used a live band playing a theme from the disco group Chic, augmented by a DJ playing short phrases on a turntable.

Only three years later, "Planet Rock" by Afrika Bambaataa in 1982, rather than being performed by professional rock or R&B players and a live drummer, used only a drum computer (known as a drum machine, here the beloved Roland TR-808), with a vocoder, a turntable, and the keyboard player John Robie playing a theme from the German electronic rock group Kraftwerk on a synthesizer. The style, which they called "electronica," sounded more rigid than the Sugar Hill style on purpose, suggesting a robotic future.

About five years later, Public Enemy welded Schaeffer's approach to Bambatta's term "hip-hop" by assembling instrumentals from looped samples taken from other records. They were able to do this without recording tape thanks to the advent of faster computers that had just become available in studios. With the producers Hank Shocklee and Eric Sadler, their looped samples provided the ominous feel of their style: 1988's "Night of the Living Baseheads" uses tiny slices from at least twenty different songs during its three minutes, including bits of Aretha Franklin and David Bowie and many from James Brown's band.

The use of stripped-down and affordable electronic instruments led to very different styles in the Bronx, Chicago, and Detroit. The Chicago house and footwork styles and Detroit techno were mostly established by musicians who had little formal training, although some, including Larry Heard, the inventor of "deep house," were virtuoso performers of jazz and pop. The instruments central to these styles were—and continue to be— turntables, samplers, synthesizers, and drum machines. The "footwork" pioneer DJ Rashad took the genre's simultaneous syncopated rhythms to a futuristic level that points out new directions for others to explore.

Leroy Jenkins, as you might imagine from his broad experience in the classical, rock, and jazz traditions, created an extraordinary range of compositions. One approach he used was to write very simple but unusual phrases as a basis for improvised longer compositions. A good start is his album *Space Minds, New Worlds, Survival of America.*

Bo Diddley stopped playing the classical violin after he heard the blues singer and guitarist John Lee Hooker; he went on to build his own trademark rectangular electric guitars and create his own style to become one of the pillars of rock 'n' roll. I played second guitar in his band, and he told me that although his daughter wanted to play the violin, he discouraged it because "there's no money in it." He nevertheless had a hit in 1959 playing a violin blues, "The Clock Strikes Twelve."

Da Hiphop Raskalz were four- to ten-year-old students I coached at the Amber Charter School in East Harlem, New York City. None played an instrument, but they immediately learned to operate synthesizers and drum machines and write their own lyrics and compositions; I only coached and mixed. Listen to "I Want Candy" by the Muffletoes and "Do the Lollipop" by Sweetness.

The performers on *Yol K'u: Inside the Sun* were Mayan-speaking students from the Seeds of Knowledge school in San Mateo, Ixtatán, high in the mountains, the first high school in that part of Guatemala. A few knew how to play traditional pieces on the giant marimbas of the region, which had originally arrived from Africa, and we coached the kids to compose their own pieces. They composed the music by placing pieces of colored tape on the marimba's various slats. Listen to "Oracion de la Cruz."

The Tangerine Awkestra were three- to seven-year-olds from Brooklyn who created original pieces on instruments they didn't know how to play during a single afternoon-long rehearsal and recording session for a full CD. The flutist and composer Katie Down and I coached them to first write a story, for which they chose a space alien invasion, and then compose music for it. They came up with brilliant avant-garde "free improvisation" compositions. Try "Aliens Took My Mom."

A profound story involving the coaching of children to create their own music comes from Conakry, Guinea, where the Canadian jazz saxophonist Sylvain Leroux has been traveling for many years to teach ensembles in a school he founded in an abandoned aircraft hanger, L'École Fula Flute. The teachers include master musicians from the Fula tradition. The students learn a chromatic version of the traditional flute, the *tambin*. Sylvain invented the chromatic *tambin* so that the instrument's repertoire could be expanded, and the students are taking off. Between their two CDs, you can hear them develop from beginners into professionals who have performed for Guinea's president: try "Bani."

1

• • • •

The Parameters of Sound

How can we measure music? How fast, long, and tall is it?

When we hear a sound, *something* must move in the air and enter our ears. These are waves in the air, which is challenging to imagine. Fortunately, ocean waves provide help.

On the beach, we see waves where the air meets the water. When you wade in the ocean, you feel the power of the wave at its highest pressure—the *peak* or *crest* that pushes you toward the shore—and you may feel pulled away from the shore at its lowest point, known as the *trough*. The ocean waves also push and pull the air around them, creating air waves that we perceive as the sound of the ocean (figure 1.1).

SIDEBAR 1.1

There is a wide range of water waves. Ripple waves, seen when the wind blows over a puddle, are short wavelength (a few centimeters), with short amplitudes, and are rapid, usually several Hz.

On California surfing beaches, the waves in a swell have wavelengths of about 150 meters, are in the range of a meter in height, and are at a much slower frequency, around one wave every ten seconds (0.1 Hz).

A giant tsunami wave caused by an underwater earthquake can have a wavelength of over 100,000 meters and be tens of meters in amplitude, with a frequency of one wave per hour (1 / 3600 seconds = 0.0003 Hz, or 0.3 thousandths of a Hz).

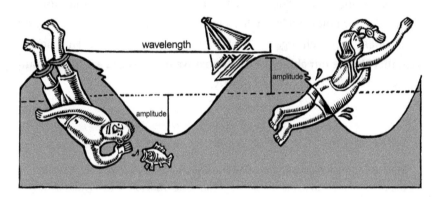

FIGURE 1.1 Waves at the beach

The waves you see and feel in the water are caused by wind, weather, and ocean currents. You also hear the sound driven by the push and pull of the water waves.

For a repeating or *periodic* wave, the *wavelength*, abbreviated as lambda (λ), is the distance between any two points where the cycle repeats. It can be measured between successive peaks or troughs.

The *amplitude* of the wave is the height measured from the midpoint of a wave to its peak or trough. We measure wavelength and amplitude in meters (*m*).

The number of waves that occur over time is the *frequency*, reported as waves per second (*s*) in units of hertz (Hz), after the physicist Heinrich Hertz (1857–1894). Two waves per second is a frequency of 2 Hz. The time it takes for a wave cycle to complete is known as the wave *period* and is the reciprocal of the frequency: the period of a 2 Hz wave is 1/2 second.

At the beach, we watch *surface* waves, where individual water molecules remain relatively local, moving in circles, up and down, or back and forth. This can be pictured by the toy boat in the figure that bobs up and down or back and forth with the waves but is neither swept to the shore nor washed out to sea: these sideways movements are known as *transverse* waves. Likewise, in sound waves, individual air particles oscillate locally back and forth between higher-pressure local peaks and lower-pressure troughs, rather than moving long distances from one end of a room to the other.

Source: Art by Lisa Haney. Used with permission.

The tide is also a wave, and as high tide occurs at 12-hour-and-25-minute intervals, its frequency is 1/44,700 seconds = 22 millionths of a Hz. If ocean waves repeated regularly (*periodically*) at a speed of at least twenty waves per second, the air waves they would produce would be heard as a musical tone. Periodic waves in air or water at 1 Hz are too slow to hear, but elephants begin to hear frequencies above 15 Hz. We begin to hear low vibrations at around 20 Hz. These very low frequencies are sometimes used in music: the low notes of a bass with a low B string "extension" is 30 Hz, and occasionally a church organ will play a very low C, at 16 Hz. That note is genuinely below our ability to hear: what we hear instead are higher multiples of the note, as explained shortly.

SIDEBAR 1.2

Musicians are familiar with these letter names for notes, but for nonmusicians: the convention is to name the notes of the scale from lower to higher pitches as A B C D E F G and then start all over again at the next higher A. The distance between the names of any of these notes and its next repeat is known as an *octave*. If this doesn't yet make sense, stick with us; it will.

At the high end of our hearing, teenagers extend up to about 20,000 Hz, above which we humans do not perceive sound. This means that with good hearing, we can hear across a ten-octave range. Dogs hear as high as 45,000 Hz——thus the dog whistle they hear but we don't——cats to about 80,000 Hz, mice to about 100,000 Hz, and some bats and whales to a ridiculous 200,000 Hz, more than three octaves higher than we can.

SIDEBAR 1.3
Abbreviations for Large and Small Numbers

Rather than writing out 200,000 Hz, we can substitute kilo (abbreviated *k*) for thousand and more easily write 200 kHz. The standard abbreviations for large and small numbers are simple to remember because they are in powers of 1000 (10^3, meaning 1 followed by three zeroes).

one thousand = 1000 = 10^3 = 1 kilo (k)

one million = 1,000,000 = $10^3 \times 10^3 = 10^6$ = 1 mega (M)

one billion = 1,000,000,000 = $10^3 \times 10^3 \times 10^3 = 10^9$ = 1 giga (G)

one thousandth = 1/1000 = $1/10^3 = 10^{-3}$ = 0.001 = 1 milli (m)

one millionth = 1/1,000,000 = $1/10^3 \times 1/10^3 = 10^{-6}$ = 0.000001

$\qquad$ = 1 micro (μ, pronounced "mu")

one billionth = 1/1,000,000,000 = $1/10^3 \times 1/10^3 \times 1/10^3 = 10^{-9}$

$\qquad$ = 0.000000001 = 1 nano (n)

The only common exception to the use of powers of three is "centi (c)" for one hundredth = 0.01, but that is generally used only for distance, measured in centimeters.

Beach Waves and Sound Waves

Water waves are helpful for understanding sound waves in the air, but there are important differences.

The peak of a water wave contains more molecules, and since liquids are hard to compress, the peak invades the airspace and becomes taller. This is known as a *transverse* wave, meaning that rather than along the long axis, in this case toward and away from the shore, the waves move in a perpendicular or sideways direction, in this case up and down. A transverse wave can also be seen in the sideways vibration of a plucked guitar string, where the string moves back and forth from where it was plucked, instead of back and forth between the ends.

Air is much more easily compressed than water. When a sound is projected from a singer or instrument, the air particles are alternatively compressed and relaxed, producing what is known as a *longitudinal wave*. For example, the movement of a stereo system speaker drives a sound wave by pushing and pulling the air in front of the speaker cone to produce regions of high (*compressed*) and low (*rarefied*) air pressure. A plucked guitar string, by vibrating sideways as a sideways transverse wave, also alternately pushes and pulls on the air around it to produce a longitudinal wave at whatever frequency it is transversally vibrating at.

These longitudinal sound waves move at the *speed of sound* (abbreviated as *c*). Instead of beach-wave-like tall peaks and short troughs, the wavelength of a sound wave can be measured by the *distance between the peaks* of compressed air. If a periodic air compression occurs more than 20 times per second (20 Hz), we begin to hear it as a note.

Sound also travels through other materials. Actually, water is still somewhat compressible, and sound under water is carried by both transverse and longitudinal waves. Solid material, like plastic and metals, are also compressible, and the sound that travels through the earth's crust also moves in both transverse and longitudinal waves.

The early physicists who wanted to understand sound waves had a challenge: how do we study invisible waves in the air? An important development was the siren, as in a police car, which was invented by the Scottish physicist John Robison (1739–1805) as a musical instrument. The siren was improved by Charles de la Tour in 1819, who named it after the mythological singing legends of ancient Greece.

De la Tour's siren works by blowing air into a tube surrounded by a spinning plate with holes bored through it. Each time the stream of air can escape through a hole, it pushes a pulse of air outward, creating a brief event of high air pressure by forcing air molecules together. The pulse moves forward at the speed of sound, in the direction from which it was emitted. As with the water molecules in an ocean wave, the air particles emitted from the siren don't actually move from the hole to the listener but force other air particles in front of it to compress and rarefy. Those particles then transmit the waves to other air particles, *propagating* a longitudinal wave (figure 1.2).

If a siren with one escape hole rotates once per second, the resulting 1 Hz air wave is a frequency too low for humans to hear, but the sound becomes audible as the hole rotates faster than twenty times a second. If you spin a siren with a single air hole at 440 rotations per second (440 Hz), you will hear the famous note known as "A 440" or "concert A," the A above middle C that an oboe plays when an orchestra tunes up.

FIGURE 1.2 A siren

A bellows pushes air through a tube, increasing the air pressure, and the high-pressure air briefly escapes through rotating holes in a disk. The frequency of the note is controlled by the number of holes and the speed of rotation: the faster the disk spins or the more holes drilled in the disk, the higher the frequency of the air waves composed of spurts of compressed air. A siren with two air holes produces a frequency twice as high as a siren with a single hole, thus producing a note that sounds one octave higher. The siren's volume is controlled by the amount of air expelled through the holes, which controls the wave amplitude.

Source: Art by Lisa Haney. Used with permission.

SIDEBAR 1.4

The note A 440 is standard for orchestra tuning in the Western Hemisphere, but some orchestras in Europe use an A above middle C (known as A4) that is slightly higher, often 442 Hz.

If we construct a siren with two holes in the rotating plate rather than one, the air escapes twice per rotation, and the pitch will be twice as high in frequency. If you spin the siren at 440 Hz, the air pulses occur twice as quickly with two escape holes (2×440 Hz = 880 Hz), producing the note A an octave higher.

In practice, musical sirens are usually made with ten or more holes arranged in a circle. For a siren with ten holes, we only need 44 spins per second to produce 10×44 Hz = 440 Hz to play the note A 440.

Now you know why the sound of a siren rises in pitch when it is first turned on and the spinning disk gains speed. Likewise, the pitch drops when the power is shut off and the speed of the spinning disk drops. (You don't yet know why its note becomes higher and then lower, like when a police car drives by. This is called the *Doppler effect*, and we'll learn about it shortly.)

As we will discuss, the frequencies of the human voice are produced in a way similar to a siren, with the glottis opening and closing to produce rapid pulses of high-pressure air.

What Range of Frequencies Is Used in Music?

When we press a key on a piano, we trigger a small hammer that hits three stretched strings, driving them to vibrate up and down in a transverse wave, like a guitar string. The vibrating strings push and pull the air around them to produce a longitudinal wave similar to that produced by the siren. The observation that a particular frequency of a vibrating string produces a particular frequency of musical note is credited to Galileo (1564–1642) and Marin Mersenne (1588–1646).

The frequency at which the piano or guitar strings vibrate results from their length, density, and tension: the longer, heavier, and looser the strings, the slower the frequency of vibration. The lowest notes of a piano have long and wide strings, and you can clearly observe their vibration, while the short, thin, high-frequency strings move too quickly for us to see more than a blur.

We use the term *fundamental frequency*, that is, the frequency at which strings or air waves vibrate, so much in this book that we will abbreviate it as f_1: the f_1 of the A above middle C (known as A4), for example, is 440 Hz. (Some books refer to the fundamental as f_0, which makes the math a little tricky. If you see f_0 somewhere else, think of it as the same frequency as f_1 in this book: this is hardly the only outmoded or confusing nomenclature in acoustics.)

SIDEBAR 1.5

Appendix 1 of this book is a table of frequencies in Hz with their corresponding musical notes. The term *middle C* arose because it was the middle note on the standard four-octave church organ keyboard. Since middle C is the fourth C counting from the bottom on a conventional eighty-eight-key piano keyboard, it is often called C4.

On a piano, the highest f_1 frequency, the C four octaves higher than middle C (known as C8), is 4186 Hz, which matches the highest note on a piccolo. These very high notes aren't commonly used in music, and even the famous piccolo high notes of "The Stars and Stripes Forever" only reach as high as F7 (2794 Hz). Gustav Mahler has the poor piccolos play a C8 in the fifth movement of his Symphony no. 2.

The strings on the lowest note of the typical eighty-eight-key piano vibrate at 27 Hz, a low A (called A0), barely over the low limit of human hearing. The ninety-seven-note Bösendorfer imperial grand piano— originally built in 1909 for the avant-garde composer Ferruccio Busoni, who wanted to play Bach organ pieces with the real frequencies—extends to the C below that (called C0), with strings so long, heavy, and loose that they vibrate at 16 Hz—you can almost distinguish the throbs of sound. This means that with a Bösendorfer, elephants hear the f_1 but we don't.

Some rock and pop performers insist on the Bösendorfer imperial grand for concerts even though they never play the low notes—and despite the fact that since the long bass strings are nine and a half feet long, the truck carrying it is too wide for the lanes on the autobahn. However, all of those strings vibrating in the low end of the piano produce a gorgeous resonance, and one luxuriates in the "sympathetic" higher-frequency harmonic vibrations produced by the unplayed strings.

Harmonics

When we play the lowest note on a Bösendorfer imperial grand, we don't perceive the low fundamental frequency but rather higher-frequency multiples, known as *harmonics*, that are within our range of hearing. As we will see later, the differences between the sounds of different instruments are largely attributable to how much of each harmonic is added. This is like tweaking the amounts of the same ingredients in a recipe.

A phenomenon you certainly experience is that the energy of low frequencies dissipates less while traveling through walls and floors than the faster frequencies of high notes. This is why you hear the bass through your apartment when your neighbors have a party and feel the bass drum in your chest at a parade. It also explains why whales and elephants produce subsonic frequencies to communicate over long distances.

Calculating the harmonic frequencies of a note made by a vibrating string of a piano or guitar is straightforward. You simply multiply the fundamental frequency by every whole number. If the fundamental frequency of a string's vibration is $f_1 = 16$ Hz,

$$\text{the second harmonic is } 2 \times f_1 = 32 \text{ Hz} = f_2,$$
$$\text{the third harmonic is } 3 \times f_1 = 48 \text{ Hz} = f_3,$$
$$\text{the fourth harmonic is } 4 \times f_1 = 64 \text{ Hz} = f_4,$$

and onward and upward, multiplying by successive whole numbers until we are beyond the highest frequency we can perceive. The harmonics of those Bösendorfer pianos must sound even more astonishing to bats.

You have now been sneakily introduced to a great deal of the math you need to understand the physics of sound and music, by realizing that a musical note has a *frequency* in the time dimension. Do sound waves also have *height* and *length* dimensions? Of course they do, or I wouldn't ask.

MATH BOX 1.1
How Fast Is a Sound Wave?

Air molecules are loosely packed and thus transmit energy and vibrations more slowly than liquids or solids. At sea level in dry air at 68 degrees Fahrenheit, sound travels at 343 m/s (meters per second), and it moves more rapidly at higher temperatures.

In water, molecules are dense, and moving in water requires much more work than moving in air, but the vibrations travel faster. The speed of sound in ocean water at 68 degrees is 1531 m/s.

In extremely dense material, vibrations are transmitted very rapidly: in a diamond, sound travels at about 12,000 m/s.

Since we know the speed of sound and that light travels much faster than sound, we can measure the distance of a lightning bolt by counting the seconds after the flash until the thunder arrives.

If we count five seconds between the lightning and thunder, our distance from the lightning is

$$5 \text{ s} \times 343 \text{ (m/s)} = 1715 \text{ m}.$$

So a five-second gap between lightning and thunder indicates that the lightning was about 1.7 kilometers away, about 1.1 miles (there are about 1.6 kilometers to the mile).

Under water, with the faster speed of sound of

$$5 \text{ s} \times 1531 \text{ (m/s)} = 7655 \text{ m},$$

thunder after five seconds indicates that the lightning bolt was 7.7 kilometers (4.8 miles) away.

SIDEBAR 1.6

Within the earth, molecules are so dense that the sound travels in "seismic waves" faster than 6000 m/s. The difference between elastic waves in the air and those in a solid is attributable to how their atoms bond and interact. In air, the particles bounce off one another like elastic balls. In a solid such as the

(continued)

SIDEBAR 1.6 (CONTINUED)

earth, atoms are more tightly bound and bounce off of one another less, so waves travel much faster. In a solid, moreover, you can also have *shear waves*, which lack a restoring force pulling the molecules back to their initial positions. Shear waves do not occur in air. Geologists record the seismic waves and use an approach similar to the comparison of the timing of lightning and thunder to determine the distance and location of earthquakes. If the vibrations of an earthquake are measured at two different points on the earth, the difference in the time when the seismic waves arrive can be used to "triangulate" and determine the precise location where the earthquake occurred.

Remember that the *amplitude* of an air wave, like a water wave, is measured from its midpoint to its bottom (trough) or peak (crest). In a sound wave in air, we measure the peak and trough of the air pressure rather than the height. The sound wave's amplitude represents the loudness, or *volume*, of sound. Loudness is typically measured in units of *decibels* (abbreviated *dB*), which means a tenth of a *bel*, named in honor of Alexander Graham Bell (1847–1922), one of the inventors of the telephone.

Our ability to perceive volume change ranges even wider than our ability to distinguish frequencies: we can perceive amplitudes that range over one million-fold.

MATH BOX 1.2
How Long Is a Sound?

We can use the rules we learned to determine not only the frequency of sound but its *wavelength*, that is, the distance between each wave's successive pressure compression peaks or rarefication troughs.

As you know, frequency (f) is expressed as events/second (s) in Hz, and wavelength (λ) is the distance between waves, expressed as meters/events (m).

If we multiply the wavelength by the frequency, the "events" units cancel, and we calculate a new value we'll call c.

$$\lambda \text{ (m/wave)} \times f \text{ (wave/s)} = c \text{ (m/s)}.$$

What is this c, which is reported in units of distance over time? Speed, of course. We already know the speed of sound waves in air or water. By rearranging this equation as

$$(c \text{ (m/s))} / (f \text{ (wave/s))} = \lambda \text{ (m/wave)},$$

we calculate the *wavelength* of a musical note.

For example, for the oboe A at 440 Hz at the speed of sound at ground level,

$$(343 \text{ m/s}) / 440 \text{ (wave/s)} = 0.78 \text{ m/wave}.$$

So the wavelength of this note is a bit less than a meter, about three feet.

In water, where the speed of sound is faster, the wavelength of A 440 is longer:

$$1531 / 440 = 3.5 \text{ m}.$$

That's about the combined height of two adults.

MATH BOX 1.3
How Loud and Quiet Do We Hear?

Sound wave amplitudes range greatly, but we don't perceive loud and quiet in a linear fashion. Sounds that are actually ten times higher in amplitude are perceived to be about "twice as loud." To measure loudness, engineers use decibels (dB) to compress the enormous range of sound amplitudes to more manageable and intuitive values.

To compress the range of these large numbers, we count the numbers of "zeroes" after 1, an operation called a "logarithm of 10" and written as $\log_{10}$. The number 1 has no zeroes, so

$$\log_{10} (1) = 0$$

(this is because $10^0 = 1$).

The number 100 has two zeroes, so

$$\log_{10} (100) = 2$$

(because $10^2 = 100$).

(continued)

MATH BOX 1.3 (CONTINUED)

The number one billion has nine zeroes, so

$$\log_{10} (1{,}000{,}000{,}000) = 9$$

(because 10^9 = one billion).

To determine how many dB louder one sound is than another, you determine the $\log_{10}$ transformation of the ratio of their amplitudes and then multiply that result by twenty.

For example, if wave A were one hundred feet high and wave B were one foot high,

$$20 \times \log_{10} (100 \text{ feet} / 1 \text{ foot})$$
$$= 20 \times \log_{10} (100),$$

and because $\log_{10} (100) = 2$,

$$= 20 \times 2$$
$$= 40 \text{ dB}.$$

This means that a hundred-fold difference in the amplitude of the two sound waves corresponds to a difference in loudness of 40 dB.

Similarly, if a tiny wave C were 1/100th of a foot tall,

$$20 \times \log_{10} (1/100) = 20 \times (-2) = -40 \text{ dB},$$

and thus wave C is 40 dB smaller than wave B.

An easy way to estimate differences in volume instantly and without calculation: when the wave amplitude changes tenfold, i.e., an order of magnitude, this corresponds to a 20 dB change.

While dB units actually measure the differences in amplitude between two waves, in common use, we provide a single number, as in "the rock concert was 110 dB." This is because the common practice is to compare the amplitude of a particular sound to a standard value of 0 dB.

The 0 dB standard chosen for loudness was defined arbitrarily as the average threshold of hearing a 1000 Hz sine wave in an otherwise silent room by young human males with excellent hearing.

This 0 dB level corresponds to a peak compressed air pressure of "20 micropascals," or "20 µPa," in which a pascal is a unit of pressure equivalent to 1 kg/m × s^2.

Thus, a sound wave ten times higher in amplitude than the 0 dB standard has a peak air pressure of

$$10 \times 20 \text{ µPa} = 200 \text{ µPa}.$$

To report this in dBs,

$$20 \times \log_{10} (200 \text{ µPa} / 20 \text{ µPa}) = 20 \times \log_{10} (10) = 20 \times 1 = 20 \text{ dB}.$$

This again demonstrates that a way to estimate the relationship between loudness and sound wave amplitude quickly is to remember that a tenfold higher amplitude than the 0 dB standard is 20 dB (200 µPa).

Now, what do these sound volume levels actually mean?

A loudness of 20 dB is close to the loudness of normal breathing or whispering heard at 2 meters distance.

A 100-fold higher amplitude (2 mPa), or 40 dB, is the background sound in a library.

A 1000-fold higher (20 mPa), or 60 dB, is the loudness of a normal conversation.

And in my favorite New York City dim sum restaurant, 88 Palace on East Broadway, which seats one thousand people, I measured 90 dB during Sunday brunch, over 30,000-fold higher than the 0 dB standard.

Prolonged exposure to sound louder than 85 dB can damage hearing, and 120 dB volumes can cause immediate hearing damage. The pain threshold is estimated as 140 dB. This sound pressure level is sadly typical of the change in air pressure driven by the speakers in a loud concert. How high is the air pressure at this concert?

$$140 \text{ dB} = 20 \times 7 \text{ dB} = 20 \times \log_{10} (10^7) \text{ dB}$$
$$= 20 \times \log_{10} (200{,}000{,}000 \text{ µPa} / 20 \text{ µPa})$$
$$= 200 \text{ Pa}.$$

Or using the rule that each 20 dB increase means a 10-fold higher amplitude,

$$140 \text{ dB} = 10 \times 10 \times 10 \times 10 \times 10 \times 10 \times 10 = 10^7,$$

a 10,000,000-fold higher amplitude than 0 dB.

This means that listeners in front of the speakers at the concert are exposing themselves to 200 pascals of sound pressure, with sound wave amplitudes 10 million times higher than the 20 micropascal (0 dB) perception threshold of someone with good hearing!

Now you have the means to measure sound and music in dimensions of speed, height, length, and frequency.

Are these all of the ways to measure music?

As an exercise, think of even more parameters. For example:

Soft to loud
Frequency of changes between loud and soft volume
Simple to complex patterns or notes
Slow to fast
Empty to dense
Groups of beats in small to large phrases
No melody to many melodies
A single harmony to many harmonies
Steady rhythm to unsteady rhythm to no rhythm
Subdividing time by very short to very long intervals
Frequencies very close or very distant from each other

This list can go for many pages and may never be complete.

An interesting challenge is to combine some of the extreme values of these different measures to create your own original musical styles.

SIDEBAR 1.7

Musical notation developed by mapping the frequency of musical notes vertically (the y-axis) and the time horizontally (the x-axis). The lines and spaces correspond to eight-note scales specified by the key signature, but for all key signatures, moving up or down by seven lines and spaces indicates an octave, or a twofold difference in Hz (figure 1.3). The x time axis changes depending on the value indicated by the shape of the note, with a whole note in 4/4 time, for example, indicating the length of time as equal to four quarter notes.

Classical musical notation is similar to the way that digital sound is displayed, which also shows the amplitude of the wave on the y-axis (measured in bits: a 16-bit compact-disc-quality (CD) file has $2^{16} = 65,536$ possible values) and time on the x-axis (a CD file is 44,100 Hz, meaning that there are 44,100 points specified per second). At the high-frequency end, the limit to encode fast frequencies is half the frequency of points, known as the Nyquist frequency (after the American engineer Harry Nyquist, 1889–1976). Thus the highest frequency a CD-type file (a WAV or AIFF) can reproduce is 44,100 Hz / 2 = 22,050 Hz.

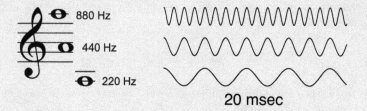

FIGURE 1.3 A440 and its octaves represented by notation and waves. Three octaves of the note A on a music stave: A5 (880 Hz), A4 (440 Hz), and A3 (220 Hz), with a 20 ms digital file showing the twofold decrease in frequency of compression/rarefaction of air pressure for each lower octave.

Source: Author.

MATH BOX 1.4
How Much Does Sound Weigh?

A strange question, and the answer is . . . tiny.

Weight is defined as mass (in kg) times the acceleration of gravity (in m/s^2), also known as *force*. While your body's mass is the same on every planet (in units of kg/human), your weight is different depending on gravity: if you "weigh 100 kg," this is because you register 100 kg on a bathroom scale calibrated for Earth's gravitation.

While we generally think of grams as a measure of weight, to physicists, weight is the force of gravity on an object and is measured in *newtons* (N), where $1 \, N = 1 \, kg \times m/s^2$.

For the acceleration of the gravity of Earth, 1 kg yields 10 N. Under Jupiter's stronger gravity, 1 kg yields 25 N, so you weigh 250 percent more, and on the moon, where 1 kg yields 1.6 N, you would weigh 84 percent less.

On Earth at sea level, a 1 cm^2 column of air from the ground to the top of the atmosphere has a mass of 1 kg. A column of air over a square meter (100 cm $\times$ 100 cm = 10,000 cm^2 = 1 m^2) has a mass of 10,000 kg. In Earth's gravity, 10,000 kg weighs 100,000 N.

We can covert newtons of weight to units of sound pressure, which, you will remember from loudness in dB, is a weight per area and measured in units of pascals (Pa).

$$1 \, N/m^2 = 1 \, Pa.$$

The air over a square meter at sea level thus has an air pressure of 100,000 Pa. This standard air pressure at sea level is also called 1 *bar* of air pressure.

Remember from our analysis of loudness that a dangerously loud sound of 140 dB corresponds to a peak of 200 Pa.

At the peak of a 140 dB sound wave, the weight of air at sea level increases only from 100,000 to 100,200 N/m^2: even an extremely loud sound weighs at its peak at most 0.2 percent of the weight of air.

So an extremely loud sound changes the mass of air from a trough of 9,980 kilograms/m^2 to a peak of 10,020 kilograms/m^2. This is indeed tiny, but because of the design of the ear, this tiny change can produce a deafening sound at the eardrum.

SIDEBAR 1.8

With the Russian satirical painters Vitaly Komar and Alex Melamid, I went to an extreme in analyzing the parameters of music. We used market research techniques to poll the public on their musical likes and dislikes. Their answers to our survey created the People's Choice Music and the "Most Wanted" and "Unwanted" songs.

"The Most Wanted Song" uses a band of the most popular instruments, consisting of guitar, piano, saxophone, bass, drums, violin, cello, and synthesizer, with low male and female vocals singing in a rock/R&B style. The most-wanted lyrics narrate a love story, and the favorite listening circumstance is at home. It is five minutes long and has a moderate pitch range, moderate tempo, and moderate to loud volume.

"The Most Unwanted Song" is over twenty-five minutes in duration, veers between loud and quiet sections and between fast and slow tempos and features timbres of extremely high and low pitch, with each extreme presented in abrupt transitions. The Most Unwanted Orchestra features the accordion and bagpipe (which tie at 13 percent as the Most Unwanted Instruments), banjo, flute, tuba, harp, organ, and synthesizer (the only instrument that appears in both the Most Wanted and Most Unwanted ensembles). An operatic soprano raps and sings atonal music, advertising jingles, political slogans, and "elevator" music, and a children's choir sings jingles and holiday songs.

The Most Unwanted subjects for lyrics are about cowboys and holidays, and the Most Unwanted ways to listen are via involuntary exposure, for example, commercials and elevator music. Statistical analysis suggests that fewer than two hundred individuals of the entire planet's population would enjoy this piece (figure 1.4).

(continued)

SIDEBAR 1.8 (CONTINUED)

FIGURE 1.4 The People's Choice: America's Most Wanted and Most Unwanted Paintings
The artists Vitaly Komar and Alex Melamid's most and least wanted paintings for the population of the United States, based on a 1993 poll of the preferences of 1001 American adults, sponsored by the Nation Institute. *The Most Wanted by Majority* was determined because 66 percent of the population preferred the color blue, 49 percent preferred outdoor scenes, 60 percent wanted paintings to be the size of a dishwasher (here, 24 × 32 inches), and 65 percent preferred inclusion of historical figures (hence George Washington). For *The Most Wanted by Minority*, sharp angle geometric forms, a small canvas (5 × 8.5 inches), and separated colors were unpopular. A similar poll for tastes in music was conducted in 1995 to produce "The Most Wanted Song and "The Most Unwanted Song."

Source: Photo by D. James Dee. Courtesy of Komar and Melamid and Ronald Feldman Gallery, New York. Copyright Vitaly Komar and Alexander Melamid.

MATH BOX 1.5

Why Do Sirens Rise and Fall in Pitch?

The Doppler effect (after Christian Doppler, 1803–1853) explains why the sound of a siren rises in pitch as a police car approaches and then descends as it drives away.

Say the siren pulses air at the A4 above middle C, 440 Hz. This corresponds to a time t between the air pulses from the siren of

$$t = 1/f = 1/440 \text{ (wave/s)} = 0.0023 \text{ s.}$$

As we learned, we can also calculate the wavelength between each siren pulse using the speed of sound (343 m/s).

$$c/f = \lambda,$$
$$(343 \text{ m/s}) / (440 \text{ wave/s}) = 0.78 \text{ m/wave.}$$

If the police car is stationary, this steady frequency is what a listener will hear. Also, if the car is moving, this is what the policeman inside the car will hear.

But if a police car is driving straight toward you with its siren blaring, the frequency you hear is higher because by the time that the siren emits its next air pulse, that pulse will have to travel a shorter distance to reach you than the one before.

If the police car is driving toward you at 100 km/h (28 m/s), by the time of the next air pulse, the siren is now closer to you by

$$0.0023 \text{ s} \times 100 \text{ (m/s)} = 0.23 \text{ m.}$$

And as long as the car drives straight toward you at the same speed, each pulse is sent from a source 230 mm closer to you. The speed of sound is the same, but the sound arrives from a closer siren.

To calculate how much the frequency is increased for the listener, add the speed (velocity) v of the car (28 m/s in this example) to the speed of sound:

$$343 \text{ m/s} + 28 \text{ m/s} = 371 \text{ m/s.}$$

(continued)

MATH BOX 1.5 (CONTINUED)

Since we now know the speed that the air pulse travels to you and that the wavelength of a 440 Hz sound is 0.78 m, we can calculate that the air pulses arrive at you at a frequency of

$$(c + v) / \lambda = f,$$
$$(371 \text{ m/s}) / (0.78 \text{ m/wave}) = 476 \text{ wave/s} = 476 \text{ Hz},$$

which is a higher note (see appendix 1), one between B♭4 and B4 above the A4.

As the car drives away from you, the air pulses arrive at you at a lower frequency, at the speed of sound *minus* the speed of the car, or

$$c - v = 343 - 28 = 315 \text{ m/s},$$
$$f = (315 \text{ m/s}) / (0.78 \text{ m/wave}) = 403 \text{ Hz},$$

between the G and G♯ below A (figure 1.5).

To calculate this more efficiently, if f_s is the sound wave's frequency when a police car and listener are stationary, and if f_m is the frequency when the

B♭ (476Hz)   A (440Hz)   A♭ (403Hz)

FIGURE 1.5 The Doppler effect

When air pulses emitted from a siren occur at a constant frequency, a policeman inside a speeding car hears a constant pitch. The puffs of air that produce the waves always travel away from the car at the same velocity, the speed of sound. But someone standing in front of the path of the car receives each subsequent siren puff emitted from a shorter distance and so hears a higher frequency. To a listener behind a siren that is driving away, each successive air puff has a longer distance to travel, so the frequency they hear decreases.

Source: Art by Lisa Haney. Used with permission.

source or listener move, we can divide them: if the siren is driving away at a velocity v,

$$f_s / f_m = (c / \lambda) / (c - v / \lambda),$$

the wavelength produced at the siren is the same and cancels

$$f_s / f_m = c / (c - v),$$
$$\text{and } f_m = (c / (c - v)) \times f_s.$$

Say you are bicycling at 10 m/s *toward* a siren producing a high A5 (880 Hz) frequency: add the speed of sound *plus* the velocity of the bicycle (343 m/s + 10 m/s = 353 m/s), and you can calculate the change in pitch as

$$(353 \text{ m/s}) / (343 \text{ m/s}) \times 880 \text{ Hz} = 906 \text{ Hz}.$$

As has likely occurred to you, one is not typically directly in front of a speeding car but (ideally) off to the side. If you are sitting in the bleachers in front of a racetrack and the cars are driving counterclockwise, they travel toward you at their maximum velocity when they are at 9 o'clock and away the fastest at 3 o'clock, and so you will hear the most rapid changes in frequencies at these positions. The driver hears the same frequency throughout.

You might imagine composing a piece performed by moving sirens—or moving listeners—traveling in a variety of directions (*vectors*), moving toward or away from one another: the pitches will be different for people at different locations.

A special case of the Doppler effect produces the sonic boom heard when airplanes surpass the speed of sound ($c = 343$ m/sec = 1235 km/hour). Here, the velocity of the plane exceeds the speed of sound c, and the sound waves emitted by its engines are very compressed. When the plane flies faster than the speed of sound, the wavelengths approach zero. Using the formula we just derived,

$$f_m = (c_m / (c_s - v)) \times f_s,$$

when the plane's velocity *is* the speed of sound, the denominator is around zero. You can't divide by zero! Small changes around zero result in the values for the frequency of the sounds coming from the plane to become very unstable, producing an enormous range of noisy frequencies that sounds like the wide frequency range of a thunderclap.

Listening #1

Very high and very low pitches are sometimes intentionally used in music to disturb the listener.

For a high-frequency note used in a classical composition, Gustav Mahler has piccolos play a C8 = 4186 Hz in the fifth movement of his Second Symphony (marker 18). It's part of a big orchestral section with lower notes played at the same time, so the piccolos are not as irritating as if they were played alone. Still, you may want to create irritating art sometime, and knowing that high frequencies can be noxious might provide some inspiration.

For examples of low frequencies used in concert music, listen to organs. In Camille Saint-Saëns's Third Symphony, known as the Organ Symphony, the end of the finale uses the lowest note on the organ, written as the C below the bass clef, which was classically called "low C" (C2). On a standard piano this would be 65 Hz, but organists on this piece use their stops to play their longest pipes, two octaves below

$$65 \text{ Hz} / (2 \times 2) = 16 \text{ Hz},$$

a fundamental frequency f_1 (C0) lower than we can perceive.

MATH BOX LISTENING 2.1

How long is the air wave of this very low C?

$$(343 \text{ meters/s}) / 16 \text{ Hz} = 21.5 \text{ meters (!)}$$

For an open organ pipe (you'll learn about open and closed pipes in chapter 3), the wavelength is twice the length of the pipe. The wavelength of middle C (C4) is about four feet (work it out), so the pipe is about two feet long (organ stops are still expressed in feet, not meters). For very low notes like 16 Hz, a cathedral must be able to accommodate a 10.75 meter open stop (about thirty-five feet).

There are two pipe organs in the world with sixty-four-foot-tall stops, the contra trombone reed stop of the Sydney Town Hall Grand Organ in Australia and the diaphone-dulzian stop of the Boardwalk Hall Auditorium organ in Atlantic City, New Jersey. Their lowest C (C1) can play a frequency of 8 Hz! To me, they sound like undersea monsters. The organist Stephen Ball has a nice demonstration of the lowest C played on the Atlantic City organ, which also happens to have the loudest stop in the world.

The Monterey Bay Aquarium Research Institute has a recording of the very low frequencies produced by an underwater earthquake. The recording requires a subwoofer speaker or headphones to reproduce.

If sound travels so well through stone that we can use it to triangulate the origin of earthquakes, shouldn't someone build instruments out of giant rocks? Watch a film of the instrument builder Pinuccio Sciola. The sounds of his sonically and sculpturally amazing instruments are barely ever heard other than on the island of Sardinia, where he made them; they are too heavy to transport.

We'll discuss low frequencies produced by elephants and whales in the last chapter. For those of us young enough to still hear it, 20,000 Hz is a high E♭10. Younger adults can usually hear 15,000 Hz, the B♭9 below the high E♭10, a high frequency currently on the border for your author. While the difference between 15 kHz and 20 kHz is only four whole notes on the piano, some sounds that define the consonant sounds in spoken language are in that range, and loss of response in that region is in part responsible for older listeners confusing words.

If music is really made from waves, shouldn't one be able to create music simply by drawing waves and then playing them back? This was done by animators in Russia in the 1920s through the 1940s. Listen and watch "Les Vautours" by Igor Boldirev and Evgeny Sholpo and anything by Nikolai Voinov using the technique that they called *paper sound.*

Music constructed from extreme parameters of quiet, slow, and long duration was a specialty of Morton Feldman: listen to a bit of his Second String Quartet, which is over five hours long.

Music constructed from extreme parameters of loud, fast, and short duration was a specialty of the hardcore band Napalm Death: listen to the deathless three-second classic "You Suffer"—but first you might lower the volume setting on your playback system.

And if you are in a safe environment with no means to hurt yourself or others, venture a few minutes of parameter-based composition in my "The Most Unwanted Song," written with the Russian conceptual artists Komar and Melamid and the lyricist Nina Mankin.

2

• • • •

The Math of Pitch, Scales, and Harmony

- How are musical scales made?
- Is it really impossible to play in tune?

Our goal for this chapter is to relate 35,000 years of history in music in twenty pages.

We know that our species perceives fundamental frequencies between 20 and 20,000 Hz. A piccolo's highest notes are under 5000 Hz, and even professional flutists find them pretty irritating. A range from 20 to 5000 Hz provides our species with a tremendous set of fundamental frequencies to create music.

A composition that takes advantage of a large number of possible frequencies within that range is Phill Niblock's *5 More String Quartets*, which uses five hundred fundamental frequencies in a single piece.

But this is extremely rare. If we exclude sliding into or out from a note, about 99.9999 percent of the music our species listens to uses very few fundamental frequencies, usually between five and twelve, along with their octave powers-of-2 multiples (this will make sense shortly).

The world *scale* means "steps" in Italian. The ordering of a small set of steps of frequencies produces a musical scale. In most cultures, the choice of a scale has mystical significance: Ling Lun was said to have developed the Chinese five-note "pentatonic" scale by cutting bamboo to lengths that when played imitated specific bird songs. Scales are nevertheless based in math, even in cases where notes are derived by what sounds "right."

The derivation of scales was classically explained by dividing a string, like a string on a guitar, and the standard approach was to use a single-string instrument known as a *monochord*. The positions that produce specific notes on a string also apply to wind instruments that produce sound by vibrating an interior column of air, like flutes and organ pipes.

Bird bones are hollow and thus good for flute manufacture—and importantly for modern musical detectives, bones survive a long time. As of this writing, the oldest known flutes are from the Upper Paleolithic period and are at least 35,000 years old. These were found in three caves in the Jura Mountains of Swabia in Germany. The caves also contain the oldest known examples of figurative art. Indeed, the flutes and artwork are ancient enough that the inhabitants possessed bones from mammoths and wooly rhinoceroses. At least one flute made from mammoth ivory has been found in these caves.

As will not surprise you, all of these prehistoric instruments are partly broken. The most complete of the flutes was discovered by Nicholas Conrad and colleagues in a cave known as Hohle Fels. It is made from a griffon vulture bone and has at least four finger holes, probably had a fifth, and it could have fit six holes. The flute was in pieces, but a replica built from a vulture bone by Wulf Hein plays a four-note scale (not counting the octave): C, D, F, B, C, followed by a higher-octave D and F. Another flute that he reconstructed plays a five-note "pentatonic" scale that remains popular today (figure 2.1).

Intact flutes have also been discovered. The oldest are nine thousand years old and made from red crowned crane bones. They were discovered by Juzhong Zhang and colleagues in Henan province in China. The best preserved of these has seven main holes and plays a six-note scale: A, B, C, D, E, F#, A. This corresponds to a pentatonic Pythagorean tuning starting on D, with one extra note. Present-day musicians still use this scale a lot!

In keeping with the mystical connotations of tuning, the math of defining scales is sometimes explained in mumbo jumbo that, in the words of Nietzsche, "muddy the water to make it appear deep." I promise that to

FIGURE 2.1 Lotta Hein plays a reconstruction of the Geissneklösterle flute
Wulf Hein's daughter Lotta plays a replica of one of the oldest flutes discovered, about 35,000 years old, discovered in 1995 in the Geissenklösterle cave in southern Germany. Constructed of a whooper swan wing radius bone, it was apparently imported, as no other swan bones were found in the cave. The flute has three holes that play a five-note pentatonic scale.
Source: Photo and flute construction by Wulf Hein, with permission of the artist.

understand why scales exist, you need only multiplication. Everything can be derived with a spreadsheet, and having a string instrument like a guitar handy will be very helpful. Hopefully, as you more thoroughly understand scales and tuning, you might also find the revelations mystical and awe inspiring.

In Western literature, the history of scales and tuning usually begins with the legendary mystic and mathematician Pythagoras (around 570–495 BCE, born in the Greek island of Samos). And why not? Pythagoras was an eccentric pacifist vegetarian searching for universal truth, and like some other inspirational figures, he left no writings but instead a tradition of interpreters who were free to reimagine and elaborate on his theories.

MATH BOX 2.1

Pythagoras's followers taught that all numbers, including musical intervals, are rational, meaning that every number can be expressed as the fractions composed from whole numbers, like 1/2 or 3/8.

Importantly for understanding musical scales, they also taught the Pythagorean theorem, which states that the length of the longest side of a right triangle (the hypotenuse) is calculated by adding the squares of the other two sides and taking the square root of the sum. For example:

$$3^2 + 4^2 = 9 + 16 = 25 = 5^2,$$

which means that the long side of this particular right triangle is the square root of $25 = 5$.

If the lengths of the shorter sides of an *isosceles* right triangle are both 1 unit, for example, $1^2 + 1^2 = 2$, then the length of the long side is the square root of 2.

Hippasus, a Pythagorean, is said to have demonstrated to outsiders that the square root of 2 could not be expressed as a simple number ratio and is therefore an *irrational* number. It is also said that this is why other followers of Pythagoras drowned him in the sea, providing an ancient warning about following charismatic professors.

The Pythagoreans thought that beans were sacred and would not eat or even step on them. Legend has it that Pythagoras was killed near his school in Croton, in Calabria, now in Italy, by an invading army from Sybaris because he refused to escape across a field of beans.

A question I ask students: if you don't have a ruler, how can you find the exact midpoint of the string?

Using a guitar or monochord, stop the string from vibrating somewhere in the middle—with your finger on a guitar or with a moveable bridge on a monochord—and pluck the string on either side. If the note is the same on both sides, *ecco*, that is half the length. On a guitar, this is at the twelfth fret.

This simple observation has given people the shivers for three millennia: half the length of the string plays a note that is twice the frequency of the open string: $2/1 \times f_1$.

In other words, turn the fraction 1/2 upside down, and you have its reciprocal, 2/1, an octave. This means that half the string *length* plays twice the *frequency*. Likewise, to play a lower octave at half the frequency, you require a string exactly twice as long.

The many tuning systems are based on this eureka moment.

MATH BOX 2.2

To find where a given note lies on a string:

In chapter 1, we found that for a wave or vibrating string

$$\lambda = c\,/\,f.$$

If you stop a string somewhere between its ends with your finger or a bridge, you produce two lengths of string, A and B:

$$\lambda_a = c\,/\,f_A,$$
$$\lambda_b = c\,/\,f_B.$$

If we divide the length of one string by the other,

$$\lambda_a\,/\,\lambda_b = (c\,/\,f_A)\,/\,(c\,/\,f_B) = (c \times f_B)\,/\,(c \times f_A).$$

The c's cancel, leaving

$$\lambda_a\,/\,\lambda_b = f_B\,/\,f_A,$$

which we rearrange as

$$\lambda_A\,/\,\lambda_B \times f_A = f_B.$$

For example, if a guitar string is tuned to A4 440 Hz and you stop the string with a bridge or finger at one-third its length (on a guitar this corresponds to the seventh fret), the frequency of the shorter side is

$$3/1 \times 440\ \text{Hz} = 1320\ \text{Hz},$$

(continued)

MATH BOX 2.2 (CONTINUED)

which is the note E6, an octave and a half higher than A4 440 Hz (see appendix 1).
The longer section of the string would vibrate at

$$3/2 \times 440 \text{ Hz} = 660 \text{ Hz},$$

the E5 an octave lower than the shorter side.

The Pythagoreans, as well as ancient Chinese and likely Egyptian musical theorists before them, were enchanted by these relationships because it meant that the notes of the scale can be calculated from rational numbers. This reinforced the idea that there must be a rational, elegant, and universal relationship of numbers in physics and art: *the universe makes sense!*

But as you have read in the story of Hippasus, this euphoric state didn't last. The vast majority of today's music uses irrational numbers, and the tension between the pure rational and the workable irrational continues to haunt us.

Octaves and Tetrachords

The relationship between two notes related by doubling (2/1) or halving (1/2) a frequency is called the *octave*: now, *octo* means eight, not two . . . but that's because the Greek scales had eight notes.

So how many notes are in scales? In Western music traditions, we mostly think of scales as consisting of eight notes:

DO, the "fundamental"
RE, the "second"
MI, the "third"
FA, the "fourth"
SOL, the "fifth"

LA, the "sixth"

TI (or SI), the "seventh"

And in the words of Oscar Hammerstein, "and that brings us back to *DO DO DO DO*," the higher-octave note that ends one octave and begins the next.

This scale in contemporary parlance is called "diatonic." Today that term does not designate what the Greeks meant by it, which was a specific tuning of an eight-note scale, but so be it. Because there are actually seven *different* scale degrees (since the eighth note is just a higher-octave version of the first note), this scale is sometimes called *heptatonic*, causing no end of confusion. In this book, in an attempt avoid even more confusion, we will count DO twice and call our diatonic scale an eight-note scale. The Greeks considered DO as a *synaphe*, a note that belongs to both the higher and lower octave.

In India around 700 BCE, the Chandogya Upanishad divided the octave into twenty-two parts, and this approach is still taught, but eight-note scales are most often used. The classical music tradition teaches the eight-note scales using the syllables SA RE GA MA PA DA NI SA. Similarly, while most Chinese melodies use only five notes, an eight-note scale is typically taught.

SIDEBAR 2.1

The note names were taken from a hymn, "Ut Queant Laxis," by Guido of Arrezzo (c. 991–1050), who is credited with inventing the musical staff for notation. The hymn praises John the Baptist and uses the first six pitches of the scale at the start of each line:

UT queant laxis

REsonare fibris

MIra gestorum

FAmuli tuorum

SOLve polluti

LAbii reatum

Sancte Iohannes

("And be loosened, resonantly sing, [of your] miraculous works, [by] your servants, loosen [our] unclean, guilty lips, St. John.")

(continued)

SIDEBAR 2.1 (CONTINUED)

The theoretician Giovanni Doni (1595–1647) is said to have modestly changed UT, which is a hard syllable to sing, to DO, which is the first syllable of *Dominus* (God) and also of his own name.

The hymn doesn't contain the seventh degree, so it simply has to be memorized. Sarah Ann Glover (1785–1867) substituted the syllable SI with TI so that the note would have a different first letter than SOL (figure 2.2).

Alternatively, DO was derived from the "D'OH" as used by the Springfield hymn poet Homer, son of Simp.

Ut Queant Laxis — Guido of Arezzo

FIGURE 2.2 UT a deer

The "Hymn to John the Baptist" by Guido of Arezzo, from which were derived the pitches and names of RE, ME, FA, SOL, and LA.

Source: Author.

There are many ways to fit eight notes into an octave scale. Much of the world's most popular music uses scales of five notes, known as pentatonic scales (yes, they should be called hexaphonic if you count DO twice), including the majority of Thai, Chinese, and Ethiopian music and a lot of blues. Composed Western music from Bach onward often uses twelve notes per octave (now *not* counting the synaphe DO twice), while the composer and instrument builder Harry Partch used forty-three notes in an octave. Each of these choices of scale degrees provides the basis for the style of the music.

Just Intonation

The word *music* is derived from the Greek goddesses the Muses, and the word *lyrics* comes from their instrument, the lyre, which provided the backup accompaniment for the lyrics of Homer and Sappho. The lyre was used to accompany singer-songwriters just as guitars, keyboards, and laptops are used now.

The lyre of the lyric poets was a harp, often of ten strings (*cords* or *chords*) spanning a bit over an octave (as one counts DO twice) and sometimes of six strings. The Greek lyre is not very different—maybe identical to—the biblical harp of David, which in the Psalms is said to have ten strings, or the contemporary Ethiopian *begena*, which also has ten strings and is said to be the harp of David brought to Africa by King Menelik I, the son of King Solomon and the Queen of Sheba, around 950 BCE. The lyre is also similar to the begena's close and popular relative, the five- or six-string Ethiopian *kraar*, which is used to accompany sung long-form poetry (figure 2.3). (The lyre is different from the *lyra*, a contemporary Greek instrument, which is a violin and accompanies lyric poetry in Crete.)

On the lyre, the frequencies between f_1 and f_2 were divided into four lower and four higher strings, known as the two *tetrachords*: the first containing DO, RE, MI, FA and the second SOL, LA, TI, and a high DO. Hence the use of the term *octave*, meaning eight notes, as it is composed of two tetrachords, even if there are only seven different note names.

The relationship between the low and high DO you already know: the string is divided in half and the frequency of the half of the string is twofold higher, the *octave* (figure 2.4).

As good Pythagoreans who find simple rational numbers spiritually satisfying, we should listen to what happens when we analyze the next whole-number fraction, dividing the string by 3. If you press on the string at 1/3 of its length (the seventh fret on the guitar), you leave 2/3 of the length free to vibrate. The resulting frequency of the longer side is the reciprocal, $(3/2) \times f_1$. So, to divide a string by a third by sound alone, use the open string as DO and find where to stop the string to hear the note SOL, the "fifth" degree. The frequency $3/2 \times f_1$ is used in virtually all tuning systems. (A rare exception is the tuning of some gamelan orchestras.)

Let's now continue in this direction and divide the string by 4. Press the string at 1/4 of its length (the fifth fret on the guitar), and 3/4 of the string vibrates at the reciprocal frequency of $(4/3) \times f_1$. This is the fourth degree of the scale, or FA.

FIGURE 2.3 Images of a begena and lyre

A muse playing a six-string lyre on Mount Helicon with a magpie at her feet (attributed to Achilles Painter, circa 334 BCE) and the contemporary Ethiopian musician Zegeye Asaye playing a six-string kraar.

Source: Vase image from Staatliche Museum Antikensammlungen, Munich. Zegeye Asaye photograph © 1995 Jack Vartoogian/Front Row Photos.

FIGURE 2.4 The monochord

A Pythagorean demonstrates how a moveable bridge produces scale degrees. Plucking the entire open string plays DO. Stopping the string at half its length and plucking either side produces the next higher DO, exactly an octave higher. Stopping the string at one-third of the length and plucking the longer side produces the fifth scale degree, SOL (corresponding to two-thirds of the full string); stopping at one-quarter the length produces the fourth scale degree, FA (three-quarters of the full string); at one-fifth the length the major third degree, MI (four-fifths of the full string); at one-sixth the full length the minor third (five-sixths of the full string); and stopping the string at one-ninth the length produces the second degree, RE (eight-ninths of the full string).

Source: Art by Lisa Haney. Used with permission.

So far, we have derived the octave, DO, by dividing by 2; the fifth degree, SOL, from dividing by 3; and the fourth degree, FA, from dividing by 4. These are classically known as the *perfect intervals* and are identical in all of the ancient tuning systems.

This still leaves the second, third, sixth, and seventh degrees, which result from smaller-fraction reciprocals of higher numbers. These are known as the *imperfect intervals*. The flavors of the different eight-note scales, particularly major and minor scales, come from these.

Divide the string by 5 (close to but not exactly at the fourth fret of the guitar), leaving 4/5 of the string to vibrate, and we have a frequency of $(5/4) \times f_1$, known as the third degree, the note MI. The frequency $(5/4) \times f_1$ is specifically a *major third*. This frequency is pretty close to but a little bit flat of what you hear on the typical piano or guitar.

Let's next divide by 6, so 5/6 of the string vibrates (close to the third fret of a guitar), with a frequency $6/5 \times f_1$, which is a smaller third degree or MI.

Ecco, you have discovered the basis for the major and minor scales! Even millennia later, the classification of a major versus minor scale is defined by the choice between the major (5/4) or minor (6/5) third. In contrast to perfect intervals, the third degrees are liquid enough that a great deal of the flavor of the blues comes from stretching where the third scale degree is placed.

These are not the only important string divisions. The division by 9 yields 9/8 = 1.125, and dividing by 10 yields 10/9 = 1.111. These are two different ways of obtaining the second degree, RE. (These are both close, but not too close, to the second fret on a guitar.)

So for the first tetrachord of a lyre, we have arrived at frequencies for each note by dividing the string length by small whole numbers. Table 2.1 shows the first tetrachord of "Didymus the musician" (around 0 CE). His tetrachord uses a major third.

TABLE 2.1

Note name	DO	RE	MI	FA
String length	1	8/9	4/5	3/4
Frequency	1	9/8	5/4	4/3

A simple way to calculate the lyre's second tetrachord, since we start at SOL, $(3/2) \times f_1$, is to repeat the process by multiplying the first tetrachord's ratios by 3/2 (table 2.2). *Voila*, a full eight-note "diatonic" scale!

TABLE 2.2

Note name	SOL	LA	SI	DO
String length	1	16/27	8/15	1/2
Calculation	$1 \times (3/2)$	$(9/8) \times (3/2)$	$(5/4) \times (3/2)$	$(4/3) \times (3/2)$
Frequency	3/2	27/16	15/8	2

There are many, many other possible variations for eight-note scales. For example, a scale advocated by Claudius Ptolemy around 100 CE differs by using 10/9 for the second degree.

This approach, using whole-number divisions of the string to derive a scale, is known as *just intonation*.

I hope that the thrill of seeing that this most fundamental building block of music can be calculated by small rational numbers produces the same satisfying glimpse into the ordered nature of the cosmos for you that it did for your intellectual and spiritual ancestors. Just intonation was the inspiration for attempts to discover "the music of the spheres," in which rationally derived musical note degrees were used to understand the structure of the universe.

SIDEBAR 2.2

There is a long effort to use musical ratios to describe the cosmos, dating at least to Pythagoras and Aristotle. In Shakespeare's *Pericles*, only the title character can hear the music of the spheres when his long-thought-dead daughter returns.

The most famous development of this concept was by Johannes Kepler, who announced that the planets rotate the sun in ellipses in his 1619 book *Harmonices Mundi*, known as *Music of the Spheres* (figure 2.5). He wrote, "We

FIGURE 2.5 Kepler's music of the planets

Each planet plays intervals defined by its elliptical orbit (in order, Saturn, Jupiter, Mars, Earth, Venus, Mercury, and the Moon). Venus, which is nearly circular, produces a drone; the Earth sings intervals of FAmine and MIsery; and Mercury rapidly climbs and descends in broad waves.

Source: From Johannes Kepler, *Harmonices Mundi*, 1619.

are to think of such circles (the orbits of the planes) as like monochord strings bent round, vibrating, and study the extents to which the parts are consonant or dissonant with the whole."

Since he discovered that planets move elliptically rather than in perfect circles, Kepler could determine the ratio between the maximum and minimum speeds during their orbit. For Earth, he determined this ratio to be 16/15, a fraction corresponding to a half step, which in the diatonic scale is also the interval between MI and FA. He wrote, "The Earth sings MI, FA, MI: you may infer even from the syllables that in this our domicile MIsery and FAmine hold sway. Venus, in contrast, produces a single note, as its orbit is nearly a circle. All of the planets might occasionally align in perfect harmony, perhaps at the time of creation."

In 2003, Andrew Fabian and colleagues, a group of cosmologists studying black holes in the Perseus galaxy cluster, discovered an enormous gas wave with a wavelength of 36,000 light-years and a frequency of 10 million years, which they calculated is a super-low B-flat, *fifty-seven* octaves below middle C. Rapidly rotating neutron stars known as pulsars emit radio waves that are often in our range of hearing, in some cases up to 400 Hz. In 2020, thousands of collaborators and the American LIGO and Italian Virgo observatories reported gravity waves associated with both stable and colliding black holes. Some gravitational wave frequencies reached 80 Hz and higher, and in a case of a collision between black holes of unequal size, they noted the presence of the third harmonic (SOL), in a way confirming the presence of intergalactic music.

Now that you understand the derivation of the ancient just-intonation eight-note scale, the assignment to you, dear reader, is to use their rules to create your own scale.

The "Circle of Fifths" and the Twelve-Note Scales

Despite the association of rational numbers with Pythagoras, the so-called Pythagorean tunings are derived in a different way, by successively dividing the string in 3 to produce a circle of fifths. The idea here is that each new frequency can in turn form its own new series of frequencies. Yet within this design lies the seed of the Pythagorean tuning's eventual destruction.

So how can you produce a scale from one fraction? As for the second tetrachord, we could multiply the fifth-degree frequency SOL, 3/2, by itself:

$$(3/2) \times (3/2) = (3/2)^2 = 9/4.$$

As notes within an octave must have values between 1 and 2, we need to bring this back to a lower octave by dividing by two:

$$9/4 \times 1/2 = 9/8.$$

Do you remember that 9/8 ratio? It is the just-intonation major third RE of Didymus the Musician.

Let's start at the frequency of RE and multiply it by SOL:

$$9/8 \times 3/2 = 27/16 = \text{the just-intonation sixth degree, LA.}$$

Once more, LA times SOL:

$$27/16 \times 3/2 = 81/32 = 2.531 \ldots$$

This number is larger than 2 and higher than an octave, so we bring it back to within the octave by dividing it in half:

$$81/32 \times 1/2 = 81/64 = 1.266 \ldots$$

Which is a ratio a tiny bit larger than $5/4 = 1.25$. Thus the Pythagorean MI is just a bit sharper than the just-intonation MI.

At this point, we have arrived at a five-note Pythagorean pentatonic scale: DO RE MI SOL LA (table 2.3).

TABLE 2.3

Note name	DO	RE	MI	SOL	LA
Frequency as fraction	1	9/8	81/64	3/2	27/16
Frequency in decimals	1	1.125	1.266	1.5	1.6875

This scale may be the most widely used in the world, and it is used in most of the popular songs from China, Thailand, Korea, and Ethiopia; many Celtic folk tunes; many blues songs; "Amazing Grace" and "Auld Lang Syne"; and a lot of American pop music, for example, the verses of Stephen Foster's "Oh Susanna" and nearly all of the Temptations' song "My Girl."

This Pythagorean tuning system can be continued ad infinitum, but something special happens at

$$3/2 \times 3/2 \times 3/2 \ldots \text{ twelve times,}$$

(which is called "3/2 to the twelfth power," written as $(3/2)^{12}$).

The number

$$(3/2)^{12} = 129.7463379 \ldots$$

is a number a tiny bit higher than 128. Why is that important?

Recall that doubling the frequency produces a note one octave higher. This means that multiplying a frequency by 4, which is $2 \times 2 = 2^2$, produces a note two octaves higher, and multiplying by $8 = 2^3$ produces a note three octaves higher, and so on. Two to the seventh power is

$$2 \times 2 \times 2 \times 2 \times 2 \times 2 \times 2 = 2^7 = 128 = 7 \text{ octaves higher.}$$

So twelve times around the cycle of fifths almost, *but not exactly,* "brings us back to DO."

This process also yields twelve notes. It is the reason why Western music uses a twelve-note scale: advancing the fifth degree twelve times *almost* completes the cycle back to DO. This twelve-note scale is called by contemporary musicians the "chromatic scale."

The "Pythagorean Comma," Irrational Numbers, and a Cosmological Crisis Where Order Breaks Down

The reputation of Pythagoras has suffered a common fate: depending on contemporary mores, he is either celebrated or blamed. The composer Tony Conrad wrote an essay, "Slapping Pythagoras," that blamed him for all of the awful styles of composition that followed from his harmonic principles, since he broke the rules both of Nature and of ethical integrity: "You traveled abroad, imperialistically raped the East of its 'Exotic' knowledge and returned with a plan to straitjacket your own people."

But thousands of years before Conrad's condemnation of this vegetarian universalist, others had identified a fatal flaw—a flaw that has haunted music theorists ever since.

$$(3/2)^{12} = 531441/4096 = 129.7463379 \ldots$$

This Pythagorean octave is not an irrational number, as it is still a product of fractions of whole numbers. But the just-intonation octave of 2/1 and this Pythagorean octave are audibly "out of tune." This inherent imperfection is called the *Pythagorean comma.*

There is an alternative way to arrive at the Pythagorean comma that derives the "whole-tone" scale. Let's take the second degree, RE, which can be the frequency $9/8 \times f_1$, and climb up six intervals, which is $(9/8)^6 \times f_1$. Now repeat this to arrive at the octave. Contemporary musicians call this the "whole-tone" scale (C, D, E, F#, G#, A#, C).

In a perfect Pythagorean world, that should yield 1 octave, or $2 \times f_1$, but

$$(9/8)^6 = 2.02729 \ldots,$$

which is technically still a rational number but not one likely to suggest that the cosmos is elegantly and entirely based on small whole numbers.

Using either cycles of seconds or fifths, the frequencies come close to but overshoot the perfect octave. And because each new frequency can initiate its own new scale-based family of division frequencies, these daughter intervals drift further and further away from the original perfect intervals.

Think about this: because Pythagorean and just-intonation frequencies are different, for any scale of twelve notes per octave, we are doomed to live with a sort of artistic and spiritual crisis, the buzz of out-of-tune notes.

What Is to Be Done?

The vast majority of music in our era is made with instruments, including the guitar, the keyboard, and the synthesizer, that use the *twelve-tone equal-tempered* tuning system.

This widespread system nevertheless frustrates musicians and listeners throughout the world. Every blues guitarist bends her strings or uses a slide to reach pitches not permitted by the instrument's frets. Every vocalist sings notes that don't match the pitches of the twelve-tone equal-tempered system.

One of the inventors of the piano, Johann Jakob Konnicke, constructed a "pianoforte for the perfect harmony" in 1796 to solve the problem of playing in tune. Konnicke used six keyboard manuals, each with its own set of strings, to provide just intonation for each key. It is said that Haydn and Beethoven played the instrument, which is now at the Museum of Musical Instruments in Vienna, but with so many strings it was difficult to build and keep in tune, not to mention perform on.

There are attempts to retain at least *some* just-intonation intervals for instruments with defined pitches. The most famous is by Andreas Werckmeister (1645–1706), who cheated—more politely called "tempered" as in tempering metal. He took four 3/2 frequencies (C to G, G to D, D to A, and B to F♯) and lowered the higher frequency by one-quarter of the Pythagorean comma. This was called "well temperament," and the tuning "Werckmeister 3" is suspected to have been advocated for by Johann Sebastian Bach in his collection *The Well-Tempered Clavier*, with includes two pieces in every major and minor key to demonstrate that well temperament could make satisfying music, even in keys with a flat $3/2 \times f_1$ (SOL). It drives proponents of alternative approaches to tuning—nowadays called "microtonalists"—to no end of exasperation when music textbooks misstate Bach's intentions by saying that this masterpiece was Bach's endorsement of twelve-tone *equal* tuning, which it certainly was not.

By far the most used approach to eliminate the Pythagorean comma is to very slightly shave down every interval in a twelve-note scale so that twelve equal steps produce a perfect octave. The idea is similar to that of the leap year. Because the year is a bit longer than 365 days, we make each fourth year 366 days, resetting the position of Earth and the Sun to where they were four years previously.

The twelve-tone equal-temperament tuning system is a surrender, an acknowledgment that if rational-number intonation systems result in a Pythagorean comma, let's at least get every octave in tune. To do so, we cheat the other pitches by shaving them down. The compromise had profound consequences for musical style. Now every step of the scale is equivalent, making it simple to jump from one key to another. This provides a basis for incredible explorations of harmony and overtone relationships within the straitjacket of the scale.

To be fair, the establishment of equal tuning was initially a radical egalitarian proposition because it used irrational numbers to calculate the notes. Yet equal tuning is also dogmatic and oppressive, and some styles of music cannot be made in equal temperament without sounding disastrously "wrong." As the political philosopher Eric Hoffer wrote in *The Temper of Our Time*, "What starts out here as a mass movement ends up as a racket, a cult, or a corporation."

For example, equal temperament does not work in Indian classical music. In this repertoire, a droned f_1 and $3/2 \times f_1$ are played throughout a piece by a *sruti box* (*sruti* means the divisions of the scale) or a *tambura*, so deviations in pitches from just intonation are heard clearly as being out of tune. As with the blues, a great deal of the art concerns how to move pitches to and away from small-number frequency ratios. The modern sitar uses frets that correspond to just-intonation spacing, and some of the frets are movable and can be slid to different frequencies depending on the scale used for the piece.

The Irrational Calculations of Equal Temperament

To understand the math of equal temperament, remember that each higher octave doubles the frequency. Selecting our f_1, DO, at the concert A4 440 Hz,

the next DO is A5 440 Hz $\times$ 2 = 880 Hz,
the next higher DO is 440 Hz $\times$ 2 $\times$ 2 = 440 $\times$ 2^2 = A6 1760 Hz,

and so on. A scale that only has one note per octave—in other words, music composed using only octaves—is easy to calculate, as each new note is $f_1 \times 2$.

Our simplest next division of the octave should be two notes per octave, exactly at the halfway point. This middle note is a frequency that when multiplied by itself arrives at the next octave.

Wuh oh. The only way to solve $x \times x = 2$ is for x = square root of 2, written as $\sqrt{2}$. Remember Hippasus's fate? This square root of 2 was the first irrational number discovered, close to 1.41421356237 . . .

The two-note equal scale is shown in table 2.4. With A4 440 as the fundamental, this halfway note falls precisely on a twelve-note equal scale's

TABLE 2.4

	DO	note between FA and SOL	DO
calculation	f_1	$f_1 \times \sqrt{2}$	$f_1 \times 2$
frequency	440 Hz	622.25 . . . Hz	880 Hz

D♮4 (622.25 . . . Hz). This dissonant-sounding and irrational interval was formerly known as the "tritone" or the "devil's interval."

A majority of baroque, classical, jazz, and popular music is *based* on finding ways to use this irrational square-root-of-2 interval to produce tension and then resolve the tension to a small real number ratio. Indeed, every jazz musician knows the harmonic sequence called "ii V," which resolves this very tension and provides the building blocks for the entire repertoire of the popular tunes of the Great American Songbook. Remember that the devil's interval, central to an extraordinary range of music, required the non-Pythagorean use of irrational numbers.

Then, to divide an octave into 3 equal intervals requires a number x for which $x \times x \times x = 2$. This number is known as the "cubed root of 2," which works out to 1.2599 . . . and is close to a major-third MI.

And jumping further ahead to twelve interval equals, the number x for which $x \times x \times x \times x \times x \times x \times x \times x \times x \times x \times x \times x = 2$, known as the "twelfth root of 2," is 1.0595 . . . This is an ungainly irrational number, but the twelfth root of 2 produces a perfectly in-tune octave, and this "twelve-equal" system is now the basis of most music in our era. It also makes some calculations easier: want to set frets for a guitar up in twelve-equal tuning? Simply make the string length to the next fret 1.0595 . . . shorter than the one before. The same goes for where to place the finger holes on a wind instrument, or the lengths of the reeds in a harmonica, or the lengths of the strings in a piano or the pipes of an organ.

MATH BOX 2.3

Equal temperament also simplifies the calculations to find the frequency of pitches. For the twelve-equal system, each higher pitch increases the frequency by the "twelfth root of 2," written as $2^{(1/12)}$. Say A4 is 440 Hz and you want to calculate the frequency of the B♭5 in the next octave higher. This is twelve steps to the next A, plus one more step to B♭, so altogether thirteen steps. You multiply 440 by 1.0595 thirteen times, that is, raise $2^{(1/12)}$ to the thirteenth power, so B♭5 is

$$(2^{(1/12)})^{13} \times 440 \text{ Hz} = 932 \text{ Hz}.$$

TABLE 2.5

	DO	RE	MI	FA	SOL	LA	TI	DO
Twelve-note equal	1	1.122	1.260	1.335	1.498	1.682	1.888	2
Just (Didymus version)	1	1.125	1.25	1.333	1.5	1.688	1.875	2
Pythagorean	1	1.125	1.266	1.352	1.5	1.688	1.898	2

How far off is twelve-equal intonation from Pythagorean or just intonation? Using the math you have learned (try it yourself), for the ratios of the eight-note major scale, the results are shown in table 2.5. So twelve equal steps are exactly correct for an octave with all three systems, but twelve equal is a tiny bit flat for the fifth (SOL) and genuinely not great for thirds (MI), sixths (LA), and sevenths (TI).

The ratio of frequencies for each of these scales may not appear too different in this chart, but can you hear if equal temperament is out of tune with just intonation? Well, yes. Let's convert these values to a major scale starting at A4 440, which we do by multiplying each ratio by 440 Hz. The octave at A5 880 Hz is perfect. SOL in twelve equal (E5, 659 Hz) is very close to its value in just intonation (660 Hz), and the difference is not noticeable. But playing the just and equal MI (C♯5, 550 and 554 Hz) together produces audible *beats* at about four times per second. (We will discuss beats in the next chapter, but you can try this now with a program that plays back sine waves.)

You may want your own music to use the genuine overtone series instead of this sloppy approximation. To help performers tune to just intonation or other gradations, we measure how far their intended notes are from the twelve-equal scale. To do this, we divide each twelve-equal step into one hundred gradations, as if there were 1200 frets per octave on a guitar.

These gradations of one step into one hundred are naturally known as *cents*. A "quarter tone," a gradation used in Middle Eastern scales that lies exactly midway between two half steps on a piano keyboard, is "50 cents" between its neighbors.

Each cent is the 1200th root of 2 (!), a number close to 1.0006: that is, multiply this number 1200 times by itself, and the product is 2.

TABLE 2.6

	DO	RE	MI	FA	SOL	LA	TI	DO
Twelve-note equal	0	200	400	500	700	900	1100	1200
Just (Didymus version)	+0	+4	−14	+8	+2	+6	−12	+0
Pythagorean	+0	+4	+8	+22	+2	+6	+9	+0

The difference in cents between twelve-equal tuning and just or Pythagorean tunings is shown in table 2.6. It is often subtle, but a chord that uses the just intervals of the genuine overtone series can have a satisfying and colorful life to it that equal temperament can't provide. Singers and string players will often adjust their intonation to produce these intervals while attempting to "sound in tune," where these intervals seem to ring.

SIDEBAR 2.3

In the cents system, every note in the twelve-equal scale degree is worth 100 cents. For example, the seventh degree of a twelve-note equal-tempered scale is defined as

$$(7/12) \times 1200 \text{ cents} = 700 \text{ cents.}$$

To determine the number of cents between any two frequencies, calculate the $\log_2$ of their ratio and multiply by 1200 cents: for example, the number of cents in the span of a just or Pythagorean fifth, a ratio of 3/2 (for illustration, A3 220 Hz and E4 330 Hz), is

$$\log_2 (330 \text{ Hz}/220 \text{ Hz}) \times 1200 \text{ cents} = \log_2 (3/2) \times 1200 \text{ cents} = 702 \text{ cents.}$$

Note that the just-intonation or Pythagorean fifth is a little sharper than the fifth in twelve-equal temperament.

Choices

In addition to the twelve-note equal temperament and just-intonation approaches, many other tuning systems are in use.

Arabic music long used a twenty-five-note unequal scale based on just intonation that is credited to the tenth-century Persian philosopher al-Farabi. As mentioned, Indian classical music in theory still teaches a twenty-two-note unequal scale based on the Chandogya Upanishad. But in practice, both the Arabic and Indian systems typically use eight-note scales, and the Indian raga system continues to use just-intonation intervals.

Contemporary Arabic and Middle Eastern music use an effective strategy to reconcile the twenty-odd-note scales with the eight-note scale. The modern Arabic tuning system, particularly in the style known as *maqam*, uses a twenty-four-note equal-tempered octave with quarter tones (that is, 50-cent intervals) that are in theory precisely halfway between the notes of the twelve-note equal scale. The adaptation of the quarter-tone scale is credited to the Lebanese theorist Mikha'il Mishaqah (1800–1888), who with his teacher modified the al-Farabi system. The quarter tones provide profound means of expression, but nevertheless, maqam scales in practice use a total of eight notes out of the system's twenty-four possible notes, similar to DO RE MI or the Indian SA RE GA.

The most popular Turkish fretted instrument, the *baglama* or *saz*, has a seventeen-note scale (the extra notes are under RE, MI, SOL, LA, and DO), with frets that can be arranged either for just intonation or in quarter tones. For example, while eight-note Western just-intonation scales used divisions of 9 or 10 for RE, the baglama has an extra in-between fret between DO and RE that corresponds to a just-intonation division by 11, as well as an additional fret between FA and SOL that can play the eleventh natural harmonic. These notes, which are required to play the baglama's repertoire, are absent from conventional Western guitars.

Those who study the practical use of quarter tones in Middle Eastern and North African traditions report that they are often not precisely in-between twelve equal half steps but are in perfect octaves with each other. When instrumentalists from outside the tradition play these tones a bit out of tune, Middle Eastern and Iranian musicians hear the music as definitively wrong and will stop the music until they are performed correctly.

SIDEBAR 2.4

Quarter-tone synthesizers are popular in Middle Eastern music: a well-known model, the Casio AT-5 Oriental keyboard, has seventeen preset microtonal scales.

In addition to twelve and twenty-four tones per octave, other equal temperaments have their advocates. There was a movement toward a seventeen-note equal scale, which happens to approximate just intervals well.

As promised, 35,000 years in a few pages. Is there an answer for how to make a scale that will please everyone? Not a chance. What will the future bring? With digital approaches we are not limited by the number of keys or holes on an instrument. But won't that new technology simply bring us back to the state of a violin or a slide whistle or slide guitar or trombone or the voice, which can sound any frequency in its range? Why, yes it will.

Consider that the sound of a lush violin section or a church choir is in large part the sound of many voices playing slightly *different* frequencies, meaning that they are a little "out of tune." Even the richness of sound and expressiveness of the piano relies on the three strings that are struck together with one hammer being a tiny bit out of tune with one another.

A violinist's or singer's vibrato is often wider than some of the small interval differences that worry some of those who care about differences in scales. My view is that it is desirable to understand the logic and consequences of tuning and scales, to use this knowledge as you prefer, and also to maintain a sense of humor about it in practice.

Now that you know how, an assignment is to invent scales of different equal temperaments or still more versions of just intonation (figure 2.6).

Then for a harder challenge, consider: all of the tuning systems discussed, including equal temperament, require 2/1 octaves and 3/2 fifths,

FIGURE 2.6 Harry Partch and his microtonal instrument ensemble

Harry Partch with some of his specially designed musical instruments that together provide a forty-three-note-per-octave just-intonation scale, from the set of *The Dreamer That Remains* (1972). The strings, blocks, and bells can be tuned to specific frequencies using principles we discuss (for malletlike instruments, jump to the design of the elephant marimba in chapter 11). For the chromelodeon organ at lower left, the reeds were cut to specific fractions marked with tape on the keys, and what would typically be a six-octave organ keyboard covers only three and a half octaves. Partch used these scales in a free manner evocative of jazz, rock, marches, and music of the Japanese Kabuki theater, with the intonations of spoken English.

Source: Photo by Betty Freeman. Copyright the Harry Partch estate.

but you *could* design a scale that lacks perfect intervals. Given that perfect intervals have been the rule for at least 35,000 years, it's unlikely that you will come up with a hit. Nevertheless, frequencies not related by octaves and fifths, as we will discuss, provide the basis for noise music.

SIDEBAR 2.5

The composer and instrument builder Harry Partch (1901–1974) read ancient Greek theory and *The Sensations of Tone* by Herman von Helmholtz and decided that equal temperament was hopelessly corrupt. His *Genesis of a*

(continued)

Music (1947) describes how he arrived at a useful forty-three-note scale and built instruments to perform it. He was further strongly influenced by the Japanese music performed in kabuki theaters, which is very effective at producing speech-like intonations and rhythms.

Perhaps the most amazing quality of Partch's music is that it so resembles the intonation and phrasing of speech that it makes one hear speech as music for hours after, the way people on the street seem like zombies after you have watched a horror movie.

Harry was notoriously obstinate, and he had to be, living at times as a traveling hobo and forging a mostly unappreciated path, although he had fans, including the composer and electronic music pioneer Otto Luening—part of the team that invented the synthesizer—who wrote the introduction to *Genesis of a Music*, and acolytes including Ben Johnston, another outstanding "microtonal" composer.

Listening #2

Let us consider how being "in and out of tune" provides mystery and beauty in sound.

It seems impossible anyone would not have heard "Do-Re-Mi," a.k.a. "Doe, a Deer," from Rodger and Hammerstein's *The Sound of Music* (co-composed by Trude Rittmann) as sung by Julie Andrews, but if you are young and inexperienced, listen once. Really, just once, or you will be sorry: you won't be able to get it out of your head.

Listen to Phill Niblock's *Five More String Quartets* to hear a rare, perhaps unique, composition that uses over five hundred different frequencies.

Ethiopian begena and kraar music, usually with a pentatonic scale and sung poetry, is probably similar to the ancient Greek styles and Hebrew styles associated with King David (refer to Psalms 33 and 137). It is used in compositions for the Ethiopian Orthodox church, known as Tewahedo. Great performers include Alemayehu Fanta and Tsehaytu Beraki on kraar and Alemu Aga on the larger and buzzier begena. You may be surprised at the vocal tone used in the singing and danceable rhythms in the Ethiopian styles, suggesting that other ancient music may have been sung with an emotion-laden gruff sound, similar to much contemporary flamenco and blues.

The ancient Egyptians used harps, as seen in their artwork, and while some of the ancient songs are still sung, the music and tunings of the harps are not clear. The Luo in Kenya (Barack Obama's father was Luo) still play a similar harp, the *nyatiti*.

Epic poetry is still sung in contemporary Greece accompanied by the lauto and lyra, particularly in Crete. The most popular epic sung in our era is the *Erotokritos*, written in the 1600s by Vikentios Koranaros. A fine singer is Nikos Xilouris. One of the top players of this repertoire is Ross Daly. Though born in Ireland, he is the acknowledged master of the Cretan tradition.

The music from Epirus in northwestern Greece still uses pentatonic scales and is thought to descend from the styles in vogue before the introduction of the diatonic scale. They have adapted the clarinet in a distinctive virtuosic style and are renowned for polyphonic choirs with one soloist and others sometimes holding drones determined by the first syllable of the stanza. One fine player who carries on the tradition of the style of the island of Chiros in New York City is Lefteris Bournias.

Listen to pentatonic blues with in-between notes by Junior Kimbrough (try "All Night Long") and pentatonic Northern Lanna Thai hill country string music known as *salah san seung*, after the names of the instruments: I recorded an album of this type of music near Lampang in 2005 (http://davesoldier.com/field.html).

Johannes Kepler's records are long out of print, but the composer/performers Willie Ruff and Laurie Spiegel have each prepared electronic renditions known as *Harmony of the Spheres* and *Kepler's Harmony of the Worlds*, respectively.

Regarding the art of the third degree (MI), namely, how just intonation in part defines music that cannot sound right in equal temperament, listen to Muddy Waters singing "Rolling Stone" (the solo version recorded by Marshall Chess of Chess records). The guitar refrain between the vocal lines moves from below, "rising" to the perfect fifth. Listen to the phrase "swimming in the deep blue sea" for an example of the blues third, which is close to the minor third 6/5, but with his bottleneck slide on the guitar, Muddy moves up a bit toward the major 5/4. Consider that this song not only gave the name to the rock band and to a magazine but that it uses only one chord and was one of the only hit records in America to feature just one singer and one instrument. (I have no idea if the composer's name relates to what Nietzsche said about muddy water earlier in the chapter.)

Then listen to John Lee Hooker's "Boogie Chillen" (1948 version), which is similar yet different. Though also a hit recorded by one musician (the taps are Hooker's feet), now the song uses two chords. And then listen to Howlin' Wolf's "Moanin' at Midnight" (the Sun Records 1951 version recorded by Sam Phillips), a hit record with one chord and an extravagant three instruments: Howlin' Wolf on harmonica and voice, Willie Johnson on guitar (who sometimes sharpens the minor SI), and Willie Steele on drums. Note the tuning on "There's SOMEbody knocking on my door" and the approach to the SOL in the hummed introduction.

A creative extension on blues tunings is La Monte Young's Forever Bad Blues Band, which used a just-intonation keyboard and Jon Calter's thirty-one-note-per-octave guitar, with the drummer Jonathan Kane providing the deep blues feel.

La Monte Young has a long-running composition at the Dream House above his apartment at 275 Church Street in Manhattan, above a pizza restaurant. The current composition, which I believe can last indefinitely, is *The Base 9:7:4 Symmetry in Prime Time When Centered Above and Below the Lowest Term Primes in the Range 288 to 224 with the Addition of 279 and 261 in Which the Half of the Symmetric Division Mapped Above and Including 288 Consists of the Powers of 2 Multiplied by the Primes Within the Ranges of 144 to 128, 72 to 64, and 36 to 32, Which Are Symmetrical to Those Primes in Lowest Terms in the Half of the Symmetric Division Mapped Below and Including 224 Within the Ranges 126 to 112, 63 to 56, and 31.5 to 28, with the Addition of 119.*

Listen to the composer/violinist Tony Conrad's *Slapping Pythagoras*. The two tracks are "Pythagoras, Refusing to Cross the Bean Field at His Back, Is Dispatched by the Democrats" and "The Heterophony of the Avenging Democrats, Outside, Cheers the Incineration of the Pythagorean Elite, Whose Shrill Harmonic Agonies Merge and Shimmer Inside Their Torched Meeting House."

To hear genuine well temperament as thought to be intended by Bach's *Well-Tempered Clavier*, there is a beautiful recording of Oscar Nagler playing the Prelude and Fugue BWV 543 on an organ tuned in Werckmeister III. Arthur Bocanneau produced a video with an electronic keyboard comparing three tunings, including equal temperament and Werckmeister III, on the first Prelude in C.

Regarding further equal-temperament divisions as well as quarter tones, listen to any piece on the Turkish baglama; a virtuoso is Tolgahan Cogulu.

A masterful use of quarter tones in Arabic orchestral music can be heard throughout the entire repertoire of Fairouz with the Rahbani Brothers, who are essentially the first family of Lebanese music. Try her heartbreakingly beautiful album of songs for Good Friday, containing the song "Ya Mariam."

For Persian quarter tones you might explore brilliant adaptations of the traditional poetic and musical repertoire from the recording *Madman of God* by Sussan Deyhim, a virtuosic singer and composer from Tehran who moved to New York and California.

Charles Ives wrote three pieces for two pianos tuned in quarter tones. This approach was extended by Ivan Wyschnedgradsky, who wrote both in quarter and eighth tones: the Arditti Quartet plays a nice version of opus 43.

The pianist Amino Belyamani has found a way to evoke Moroccan microtones with the jazz piano in his group SSAHHA. He detunes two notes per octave, the E and B, by about a third tone, that is, 33 cents, where 100 cents equals a half step. The concept is not too different from well temperament.

There are an enormous number of modern compositions for alternative tuning systems, and a resource is the American Festival of Microtonal Music run by Johnny Reinhard. Personal favorites of mine are Ben Johnston's string quartets.

With computers and synthesizers, it is far more straightforward to produce precise pitches that change depending on their relationships to other pitches, and the primary innovator of this is Wendy Carlos, also a synthesizer pioneer. Listen to *Beauty in the Beast*. On this recording she introduced "alpha" and "beta" scales that produce good triadic chords but do not use octaves.

Jacob Collier creates arrangements of rich vocal harmonies in the tradition of the Four Freshmen, Beach Boys, and Take 6, but his are sometimes sung in just intonation. The logic of the overtone series can lead to key centers quite distant from ones playable on the piano; for instance, if one modulates from C major to the key of its just-intonation eleventh, we are in a scale with a fundamental that is a quarter tone between F and F♯: in his work, this sort of approach sounds natural and lovely. Try "In the Bleak Midwinter."

You can listen to anything by Harry Partch, and there is a film, *The Outsider*, of him playing his instruments at home. For me, Partch comes the closest to expressing spoken English on musical instruments, and after listening to his music, you'll hear everything in a different way. Watch the astonishing University of Illinois 1961 version of "Revelation in the Courthouse Park," featuring the god Dionysus paired with the contemporary rock 'n' roll god Dion (who must have been named with awareness of Dion DiMucci from the fantastic Bronx do-wop singing group Dion and the Belmonts). This opera combines elements of Athenian and Japanese kabuki theater, John Phillip Sousa, Charles Ives, Busby Berkeley, big-band jazz, circus bands, Rodgers and Hammerstein's *Oklahoma*, barbershop quartets, and Chuck Berry–era rock 'n' roll.

Partch's theater music is strongly influenced by the Japanese kabuki tradition, including the rhythms and the intonation of the vocals and certainly the sounds of the instruments: one great album is the 1954 *Azuma Kabuki Musicians*, directed by Katsutoji Kineya and Rosen Tosha, performing a program of nagauta music. Kabuki was originally performed by women, but the tradition transformed to be performed exclusively by men, who sing the female characters.

3

• • • •

Waves and Harmonics

- What are sine waves, and how are they components of sound?
- What are harmonics, and how do they produce the sounds of musical instruments?

One of our astonishing abilities is that we can identify the source of specific sounds in complex sonic environments, for example, one speaker among multiple conversations at a party (the "cocktail party problem," named by the psychologist Colin Cherry in 1953), or follow an individual instrument within a big band.

To realize how impressive this is, consider that we can record a party or orchestra with a single microphone that transduces the changes in the pressure into a single wave at the microphone membrane. Then we can recreate this same sound wave in another room simply by vibrating the cone of a single speaker in air, and we can still identify the original sound sources that, added together, produced that single wave.

This feat requires our ears to act like microphones that pick up a complex wave and a nervous system that isolates and identifies that wave's original components. How we manage to do this is the topic of most of the rest of this book.

Circles, Strings, and Clocks

Let's begin with the simplest wave, a circle.

Understanding the properties of the *sine* wave, the wave inscribed by a circle, may seem intimidating, harking back to the terror of high school math class. Simply remember that we are talking about circles and that the word *sine* was not designed to frighten artists but was a synonym for the friendlier-sounding word *string*.

> **SIDEBAR 3.1**
>
> The word *sine* came from a "string" of translation errors. The Greek word for string, *khorde* (chord), was said to be translated into the Sanskrit word *jya*, the string of a bow and arrow. *Jya* was translated into Arabic as *jiba*, which was mistaken by other translators as *jayb*, meaning a "fold" or "bosom." This was then translated from the classic book on algebra, written in the 800s by the Persian Al-Kwharizmi (whose name is the basis of "algorithm"), by the English Arabist Robert of Chester in the 1100s into the Latin word for bow or bosom, *sinus*.

Making a sine wave is easy. Tie a string to a post in the ground, pull the string taut, and walk around the post. You are walking in a circle with a *radius* that equals the length of the string. The time it takes you to walk the circle is the *period*. The distance you walk around the circle is the *circumference*. And you, dear reader, are the *sine wave*.

To draw the sine wave over time, which is how sounds are stored after they are recorded with microphones to be later played back through speakers, consider a clock with an hour hand. The clock hand makes a wave with a peak/crest at twelve o'clock (known as both 0 and 360 degrees), a trough at six o'clock (180 degrees), and two midpoints at three and nine o'clock (90 and 270 degrees). The period of this sine wave from peak to peak for the hour hand is twelve hours. If we plot the position of the hour hand on the y-axis and the time along the x-axis, we draw a sine wave (figure 3.1).

To read the *sine of the angle* of the clock hand as it rotates, think of the distance from a vertical midline running between twelve and six o'clock

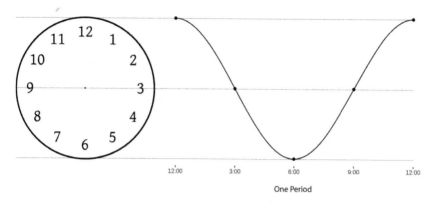

FIGURE 3.1 Clocks produce sine waves

The rotations of a clock hand produce sine waves with twelve o'clock the peak, six o'clock the trough, and a midline at three and nine o'clock.

The *period* of the wave (use the term *period* when measuring a cycle over time and *wavelength* when measuring over distance) is measured as the duration of time between each peak, or between each trough, or between each incline or descent that crosses the midline. In each case, the *period* of once around the clock is 12 hours, a very slow frequency wave.

2 (rotations/day) / 86,400 (seconds /day) = 0.000023 Hz = 23 μHz.

The *amplitude* of the wave is the height of the clock hand, and the *radius* of the circle is the clock hand length to the circle's edge: both are measured in meters. The circumference, the distance along the edge of the circle, is also measured in meters and is equal to 2 × the radius × π (about 3.14).

Source: Art by Jai Jeffryes. Used with permission.

to wherever the clock hand is. At twelve o'clock (0 and 360 degrees) and six o'clock (180 degrees), the clock hand is exactly on that midline, so the distance is zero. Therefore, the sine of 0, 180, and 360 degrees is zero. At 3 o'clock (90 degrees) or 9 o'clock (270 degrees), the hand is the furthest distance from the vertical midline, so the sine of 90 degrees = 1 and at 9 o'clock (270 degrees) is –1. This means that the sine of a clock runs between 0 at noon and midnight, has a value of +1 at three o'clock, returns to a value of 0 at six o'clock, and has a value of –1 at nine o'clock.

MATH BOX 3.1

To measure cosines, we do the same calculation, except we measure from a *horizontal* midline running between 9 and 3 o'clock. The cosine of three o'clock

(continued)

Waves in the Real World Have Harmonics

In the real world, not only do clocks produce sine waves, but so do springs that move up and down or back and forth. And so does the cone of a stereo speaker. If a speaker reproduces a recording of a "pure tone," say concert A 440, its cone moves back and forth, pushing and pulling on the air, to produce a longitudinal sine wave at a rate of 440 Hz.

If you listen to a sine wave through a speaker, you will realize why pure sine waves are rarely used in music: they're pretty boring. As pointed out in 1843 by Georg Ohm, whose work on electrical currents we will discuss later, nearly all of the musical sounds we use are constructed from combining a fundamental frequency f_1 and its harmonics, which you will remember from chapter 1 are small real number multiples of f_1.

We can hear harmonics clearly on a monochord or guitar by slightly modifying the way we derived just intonation from dividing the string's length by small numbers. The difference is that to hear harmonics, rather than stopping the string entirely as with a bridge, instead just place your finger lightly on the string, to dampen it from vibrating at that point.

If you place your finger lightly to dampen the string at the precise half-length of a string (for example, the twelfth fret of the guitar) and pluck either side, you hear the *first harmonic*, f_2, a frequency exactly twice as high as the fundamental note, f_1, of the open string. This is because only the two halves of the string can vibrate, and, as you know, half of the length of the string sounds an octave higher.

The points where the string cannot vibrate, either because it is attached at the ends or because you have dampened the vibration with your finger in the middle, are called *nodes*. The midpoint of the two vibrating portions, where the amplitude is greatest, is called the *antinode*.

SIDEBAR 3.2

The use of nodes to describe string vibration was introduced by the English physicist John William Strutt, 3rd Baron Rayleigh (1842–1919), who conducted his experiments in his home laboratory. He published *The Theory of Sound* in 1877, a book still consulted on the physical aspects of acoustics.

If we lightly stop the string at one-third of its length (the seventh fret on a guitar), the string vibrates in three equal parts, with one node where your finger dampens the string and a second node that forms at the precise middle of the longer side. The pitch from the one-third lengths of strings is three times the frequency and is the *second harmonic*. As you know from the scale derivations, the frequency that is 3/2 times the fundamental is a perfect fifth (SOL). Three times the fundamental frequency of the string produces a fifth in the next octave.

As with just intonation, we can now continue to dampen the string at higher-number divisions, each of which produces a smaller length of string and a higher frequency. With a string length of 1 as the fundamental f_1,

1/2 yields $2 \times f_1$ and the second harmonic (f_2), an octave,
1/3 yields $3 \times f_1$, which is the third harmonic (f_3), an octave and a fifth,
1/4 yields $4 \times f_1$, which is the fourth harmonic, two octaves higher (because $4 = 2^2$),
1/5 yields $5 \times f_1$, which is the fifth harmonic (the just-intonation major third in the second octave),

ad infinitum, with closer and closer intervals with more divisions (figure 3.2).

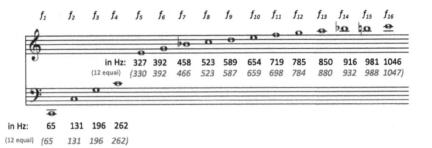

FIGURE 3.2 The overtone series

A fundamental f_1 of low C (C2: 65.4 Hz) and its first 16 harmonics. These are easy to calculate. Simply add 65.4 Hz for each subsequent step. The upper frequencies are in just intonation and indicate the components of the sound of most string and wind instruments and the voice. The closest notes in standard twelve-equal tuning are shown underneath in italics (refer to appendix 1). Some of the harmonics are quite far from equal temperament, especially the seventh, eleventh, and fourteenth. The notes on horn instruments such as the French horn and trumpet are played using the harmonic series, and some trumpet virtuosos play higher than the sixteenth harmonic.

Source: Author.

Since harmonic frequencies are in just intonation, instruments such as the French horn that use the overtone series must compensate to play closer to the equal-tempered tuning of other instruments, which is a particular challenge when dealing with the seventh, eleventh, and fourteenth harmonics.

While the sounds that musical instruments produce are made up of the harmonic series, the higher harmonics typically represent smaller and smaller components of the overall sound. Indeed, in professional recordings the very high harmonics tend to be filtered out on purpose. This is a double-edged sword: too much high end can sound like extra garbage, but too little muffles the sound.

Making Harmonics Visible

It may be easier to accept that harmonics in vibrating strings divide themselves into equal parts if you see it. One way to observe nodes and harmonics is with a jump rope, which also divides itself into equal parts.

The paradigmatic harmonic demonstration toy is the Slinky, a spring of thin wound steel or plastic that currently sells in New York City for a dollar. The coils provide a way to compress a longer string into a shorter distance.

If you and a friend hold the two ends of Slinky and move it back and forth to simulate a jump rope, you two are the wave nodes, while the middle, which moves the greatest distance, is the antinode, and you will see the "first harmonic." If you increase the velocity at which you and your friend swing the Slinky, you can create a new node in the middle, producing the second harmonic, and then with some effort, you can even produce two nodes, the third harmonic. If you are fast and regular enough in your oscillations you would hear the harmonic notes from this large vibrating string (figure 3.3).

When used in this fashion, the Slinky is forming transverse *standing* (a.k.a. *stationary*) waves; that is, the nodes and antinodes remain at the same spot (we discussed these waves in chapter 1). Similarly, a plucked guitar string has major stationary nodes (the bridge and nut) and antinodes (the middle of the string) and also has nodes and antinodes at the harmonic positions.

Standing sound waves are not only present in strings. As we will see, they also form within wind instruments like a flute or an organ pipe, in which they form not in a vibrating wire but in the air.

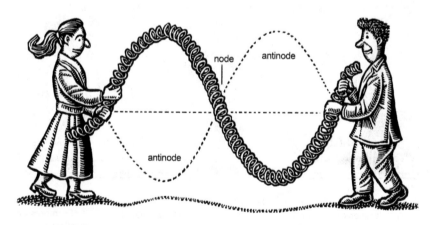

FIGURE 3.3 Making the second harmonic with a Slinky
When two people wave a Slinky slowly to the right and left or up and down like a jump rope, the midpoint shows the maximum crest and trough (the vibrational antinode), with the two held ends forming the nodes of a transverse standing wave. With increased speed, a new node forms in the middle with two antinodes, and the Slinky displays the second harmonic, as in this drawing. With skill—try working the wrist into it—you can form two more nodes, dividing the Slinky into three, for the third harmonic, and into four, for a fourth harmonic.
Source: Art by Lisa Haney. Used with permission.

An exemplary demonstration of a stationary sound wave is Kundt's tube, after August Kundt (1839–1894). A transparent tube contains talcum powder or small foam pellets. When a speaker is attached to one end, if it plays a frequency with a wavelength that matches the length of the tube or a harmonic of the wavelength, the sound becomes louder as the entire structure resonates between the two ends. The wave appears to stand still as the powder gathers at the nodes, eerily suspended in the air by sound.

For a dramatic and larger version of a standing wave in a Kundt tube, set up a speaker at one end of a hallway, producing a longitudinal traveling wave. At some low frequencies, the sound reflects back from the walls and the waves moving in opposite directions reinforce (called *constructive interference*) and cancel (called *destructive interference*) each other to produce stationary nodes and antinodes. As one walks through the hallway, you can hear loud antinodes where the air density compresses and rarefies and quiet nodes where the air pressure does not change (figure 3.4).

FIGURE 3.4 Producing a standing wave in a hallway

The vibrations of the speaker cone set up a low-frequency longitudinal wave of compressed and rarefied air pressure. At the rear wall, the air particles are restricted and reflect back to form an air pressure antinode. This reflected wave moves in the opposite direction and sums with the speaker wave to form standing waves with stationary antinodes and nodes. As you walk through the hall, you will hear volume swells at the pressure antinodes and diminuendos at the nodes where the pressure is constant. Changing the frequency will change the positions of the quiet nodes and loud antinodes.

Source: Art by Lisa Haney. Used with permission.

Wind Instruments with Open and Closed Ends

The sound of wind instruments such as trumpets, saxophones, and organ pipes are produced by standing waves that form inside the instruments.

For instruments like the clarinet, the player blocks the air they blow from escaping back through the mouthpiece. As the air particles can't escape, the closed mouthpiece end maintains a high air pressure and thus the lowest vibrational motion of air particles. The open "bell" end is exposed to the lowest air pressure, the ambient air in the room, so there is a maximum range of motion for the air particles, and they alternate between denser and rarefied states.

Since the position of the vibrational antinode is defined by the distance between the open and closed end, the wavelength of these instruments is controlled by the length of the tube. To produce a higher note on a wind instrument, for example, a flute, clarinet, or saxophone, one essentially shortens the distance/wavelength by opening a hole for air to escape before it reaches the end of the tube. Alternatively, the distance and wavelength can be made longer by increasing the length of the tube, for example, by using the valves of a trumpet or slide of a trombone.

For example, figure 3.5 shows that a full wavelength for a clarinet, which has one open and one closed end, is four times the length of the tube. The B♭ clarinet is 60 cm (0.6 meters) long. Therefore, to calculate the wavelength, multiply by four:

$$\lambda = 4 \times 0.6 \text{ m} = 2.4 \text{ m}.$$

Thus, a clarinet f_1 at sea level should be

$$f_1 \text{ (wave/s)} = (343 \text{ m/s}) / (2.4 \text{ m/wave}) = 143 \text{ Hz},$$

which (see appendix 1) is D3 below middle C4, and that is indeed the lowest note playable on this clarinet.

Next, let's consider the fundamental frequency f_1 of a trumpet, which also has an open and closed end. A stretched-out conventional B♭

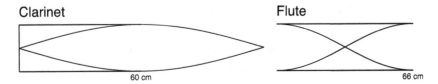

Clarinet

Flute

60 cm

66 cm

FIGURE 3.5 Sound waves formed in tubes

At left, with a clarinet, the player's mouth and positive pressure block air from escaping from the mouthpiece, and so we draw one closed and one open end. The mouthpiece is a node of highest air pressure and lowest air particle movement. The open end has the greatest air particle movement between peaks (*compression*) and troughs (*rarefaction*). To form the next node in the wave after a peak at the open end requires a full length of the clarinet in front of the instrument, as does forming the next node after a trough: therefore, a full wavelength of a clarinet standing wave is four times the length of the body.

At right, a flute is represented by a tube with two open ends. (A standard flute has one genuine open end, but air also escapes through the mouthpiece, which is only barely covered by the player, and so emits air through two "ends.") This means that both "ends" act as anti-nodes where the air particles vibrate between air density peaks and troughs. The middle of the flute body is a node of the highest pressure and the lowest air particle movement. Thus, the flute alternates between antinode peaks and troughs at both ends, and the full standing wave is twice as long as the flute body.

A clarinet and flute are nearly the same length, but as the wavelength is four times the length of the clarinet and twice the length of the flute, the clarinet's range is about an octave lower.

Source: Art by Jai Jeffryes.

trumpet tube without the valves depressed is 1.48 meters, so the wavelength is

$$\lambda = 4 \times 1.48 \text{ m} = 5.92 \text{ meters,}$$

and

$$f_1 \text{ (wave/s)} = c \text{ (m/s)} / \lambda \text{ (m/wave)} = (343 \text{ m/s}) / 2.96 \text{ m/wave} = 58 \text{ Hz.}$$

This is a pretty low pitch, the note B♭1, three octaves below middle C. This fundamental note is rarely if ever played, though, possibly in growls in New Orleans–style playing and by the Cuban American virtuoso Arturo Sandoval.

Before the addition of keys and valves to trumpet design, "natural trumpets" would play notes starting from the fourth harmonic up to the sixteenth harmonic. Remember that $4 \times f_1$, the fourth harmonic f_4, is two octaves higher than f_1 (because $4 = 2^2$):

$$4 \times 58 \text{ Hz} = 232 \text{ Hz},$$

which is the B♭3 below middle C4 (see appendix 1). This is normally the lowest performable note for a trumpet without valves.

On the standard B♭ trumpet, if one opens all three valves, the tube is lengthened to about 1.77 meters, lowering the f_4 to a low E3 (165 Hz). With this contemporary valve setup, modern trumpeters can play each twelve-note pitch starting from this note.

This situation is different for a flute. The concert flute is 66 cm long (0.66 m), a little longer than the clarinet, but instead of producing a fundamental frequency lower than the clarinet, its low note is middle C, nearly an octave higher.

A modern flute has one open end, and the other end is stopped with a cork. As always, the player applies air through the mouthpiece to produce a vibration in the tube. But in this case, she doesn't actually cover the mouthpiece as do horn or reed players, so the air also escapes through the mouthpiece; thus there are two "open" ends. With this design, the node of highest pressure and lowest vibration is in the middle of the tube.

The vibration between the nodes of compressed air and antinodes of rarefied air now occurs between the two open ends. With two open ends, the frequency of vibrations between the peaks and troughs in the flute is twice as rapid as for the closed-end trumpet or clarinet, and the f_1 frequency therefore corresponds to a wavelength twice the length of the tube.

Thus, f_1 for a flute at sea level is

$$f_1 \text{ (wave/s)} = (343 \text{ m/s}) / (2 \times 0.66 \text{ m/wave}) = 260 \text{ Hz},$$

which is the note middle C4, the lowest note played by the conventional flute.

My favorite way to demonstrate the change in the fundamental frequency of a tube with one versus two open ends is with a poster tube. If you bounce an end on the floor with both ends open, you set up a wave with a node in the middle like a flute and hear a pleasant sounding f_1, like with a flute. If you stop one end and bounce again, the internal node disappears and you produce a node in the air at twice the height of the tube, and you hear an f_1 an octave lower, like with the clarinet.

Organ builders know the poster-tube trick, and church organs use both open- and closed-pipe pitches. As mentioned in chapter 1, open organ pipes must be 21.5 meters high to produce the lowest C notes used in their conventional repertoire, but if you close the pipe on one end, it only needs to be half as high.

SIDEBAR 3.3

The standing waves produced in flutes and organs have an unappreciated role in the foundation of modern physics and technology. In his article "The Fundamentals of Theoretical Physics" (1940), Albert Einstein wrote that Louis De Broglie was considering the problem of why energy wavelengths at the atomic level appeared to exist in jumps, or *quanta*, rather than in continuous values. Inspired by the way that stationary waves in flutes and organs form discrete wavelengths (corresponding to $f_1, f_2, f_3 \ldots$) based on the position of their nodes, De Broglie suggested that quanta could be explained if electrons orbit the atom at similar discrete levels, thus introducing the basis of atomic structure. The quantum mechanisms inspired by the working of musical instruments underlies transistors and computers, lasers, and a host of everyday tools.

Seashells, Horns, Trumpets, and Standing Waves

The instruments of the horn family are based on animal horns. The *shofar* played during Jewish religious services is still the horn of a ram, or for the Jews from Yemen, the horn of a greater kudu, which has a twisted shape that generates a sweeter and haunting tone.

Animal horns, usually with a finger hole to change the effective length of the tube, are used as musical instruments throughout Africa, including in religious ceremonies and funerals. An instrument named after the traditional Zulu *vuvuzela*, based on an antelope horn, is now made from plastic or metal and used to play obnoxious just-intonation intervals by fans during soccer matches.

Metal horns have been produced from at least 500 BCE, as seen from excavations of bronze instruments in Scandinavia now at the National Museum in Copenhagen. Likely used for religious ceremonies, the *lur* was two to three meters long and shaped like the letter *S*, with a cup-shaped mouthpiece and a conical bore like the modern horn, but the instruments were made in pairs that twisted in opposite directions, with the bells high above the player's head.

SIDEBAR 3.4

The shell of the conch is used as a musical horn in India, the Caribbean, and the Pacific islands. The jazz trombone player Steve Turre is a virtuoso on the conch shell, and by cupping his hand in the shell to change the length of the vibrating air, he seems to be able to play any note he chooses.

In a horn, one end of the tube is open to the world, and the other has a player puffing air into a small mouthpiece; that is, there is one open and one closed end. One would think that the fundamental f_1 would be four times the length of the tube, as with a clarinet, but horns instead behave like flutes, as if they had two open ends.

Horns, and the French horn in particular, have a narrow hollow end at the mouthpiece that gradually flares out toward the bell. This shape is a cone, and the airwaves it produces are different from those produced in a nearly straight cylinder like a clarinet, with the reflections of air along the tube's walls producing waves of spherical shape.

For French horns, the natural trumpet, animal horns, shells, hunting horns, and the bugle, the harmonic series provides every note available to the player. For classical orchestras from Bach to Beethoven, players would bring a collection of horns and trumpets with different tube lengths (*crooks*) and fundamentals and choose the one specified by the key of the piece.

This means that horns are instruments that play in genuine just intonation, and for them to work well with an equal-tempered orchestra requires specialized skill.

The reason that conical horns act as if they had two open ends is a subject of many argumentative analyses. In practice, the air at the tapered end at the mouthpiece and the bell are both vibrational-motion antinodes with fundamental frequencies about twice the length of the instruments, and they produce a full harmonic series in the same pattern as for the flute.

For example, a standard French horn uncoiled with the valves depressed is about four meters long, and a tube with two open ends would produce a fundamental of

$$\lambda = 2 \times 4 \text{ m} = 8 \text{ m}.$$

Thus, f_1 at sea level should be

$$f_1 \text{ (wave/s)} = (343 \text{ m/s}) / (8 \text{ m/wave}) = 42 \text{ Hz},$$

about F1, which is the lowest playable pitch and one that is sometimes used in the orchestral repertoire (where it is transposed to help reading).

On the second, or f_2, harmonic, the brass players essentially divide the tube into two parts by forming a node in the middle of the tube, like the harmonics of the string or a Slinky, to produce the octave. As both ends form antinodes like a flute, the entire wave for the second harmonic f_2 is the full length of the tube.

A horn player divides her tube into three equal parts for the f_3 harmonic, four equal parts for the fourth harmonic, and on to an extreme number of parts by varying the air pressure and the shape of her mouth and lips, called the *embouchure*. A tighter pursing of the lips results in higher pressure, which slices the tube of air into multiple geometric divisions, like when the Slinky is vibrated faster and faster. What's more, brass players can produce a wide range of pitches without any horn at all, using only their mouthpiece, breath, and embouchure to get the mouthpiece to vibrate at a specific frequency.

Once a horn player has the skill to navigate between the eighth and sixteenth harmonics on the actual horn, three octaves above f_1, they can play a nine-note scale, including all of the pitches of an eight-note just-intonation scale, although the 1/11 division f_{11} is far too sharp for the 4/3 FA. The 1/7 division f_7 is so far from the equal-temperament seventh that modern brass players are forced to find alternative ways to play the note, even though it is the *rest* of the world that is actually wrong! French horn players insert and open their hand into the bell to shorten the wavelength, although when they do this they typically switch to a higher harmonic and so in practice they use this technique to play lower frequencies.

With the trombone, the slide changes the brass tube length to anywhere the player can reach, and these microtonal glides between scale notes produce a trademark of the New Orleans brass tradition.

Ocarinas and Resonance Chambers

The ocarina, a potato-shaped wind instrument used by the Mayans and Aztecs and in India, Italy, and China, cannot work in the same way as flutes, clarinet, or horns, as it is only a few centimeters in length and produces pitches far lower than those corresponding to sound waves four times the length of the instrument. Ocarinas, soda bottles, and seashells cupped over your ear are instead *Helmholtz resonators*, after Hermann von Helmholtz (1821–1894).

The ocarina has a bulbous chamber and a lip known as a fipple into which the player blows. When the player blows, the air pressure becomes greater inside the fipple, and air particles move back and forth with alternating compression and rarefaction between the walls of the chamber, like a spring. Lifting the finger to uncover holes in the ocarina body decreases the air pressure within the chamber so that the frequency of vibration is higher.

Beyond ocarinas and soda bottles, resonators are central to the sound of instruments like guitars and cellos. This is obvious if you have played a string on an acoustic guitar, which resonates the air inside the box, and then the same a string on a solid-body electric guitar with no air inside to resonate. The acoustic guitar produces a rich sound and the electric only a tinny twang. One can feel a cello's or acoustic guitar's body vibrate as it expands and contracts at characteristic frequencies.

One can also see that these resonances are composed of standing waves using Chladni figures (after Ernst Chladni, 1756–1827). Take a thin metal pan or plate, coat it with a little flour, rice pellets, or sand, and then vibrate the plate by stroking it with a violin bow or by playing a note through a speaker cone to vibrate the plate bottom. The pan creates nodes of low vibration where the rice grains gather and are not bounced out of place, forming beautiful geometric wave patterns. The nodes and antinodes on a musical instrument's body and on other vibrating surfaces can be identified in this way (figure 3.6).

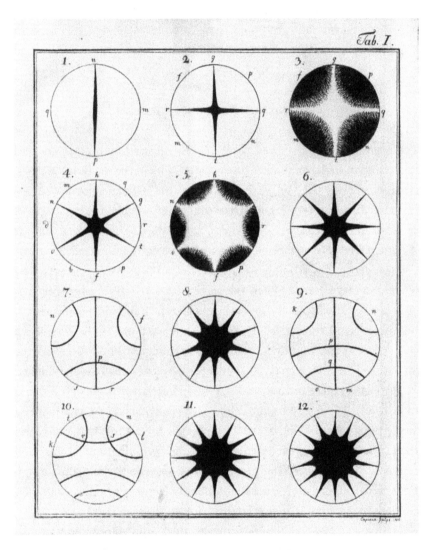

FIGURE 3.6 Chladni figures
Rice pellets gather in vibrational nodes and are bounced out of vibrational antinodes, producing a visualization of the vibrational patterns of guitar and violin bodies, sheets of metal, and other resonant structures.

Source: Ernst Chladni, *Entdeckungen über die Theorie des Klanges*, 1787.

The resonances can also be heard by tapping the instrument's body without the strings. String instrument makers typically try to make them resonate at about the frequency of their second lowest string, at the A2 string on a guitar, a D4 on a violin, G2 on a cello, and A1 on a bass (figure 3.7).

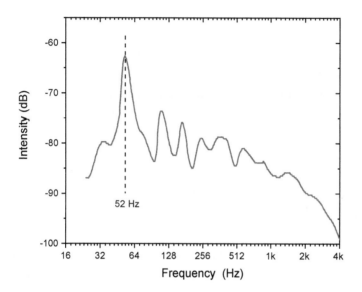

FIGURE 3.7 Double bass resonance

A frequency spectrum recorded when tapping the front of a very well made eighteenth-century Italian orchestra double bass without its strings. The maximum peak resonance frequency is about 52 Hz, corresponding to the fundamental frequency f_1 of the open second-lowest string, A1 (55 Hz standard pitch in the United States, see appendix 1).

If the instrument were strung, bowing or plucking would vibrate the bass and reinforce these frequencies at these peaks, providing a far richer sound than if the string was attached to a slab of wood, as with a solid-body electric guitar.

Source: Recording by Giancarlo Ruocco of a bass crafted by the Carcassi family workshop in the 1700s at David Gage Instruments, New York City. Used with permission.

Our mouths are also resonators, and the characteristic resonant frequency of an individual's mouth provides the distinct identity of different singing and speaking voices.

The control of the resonance of our mouths is used to produce the pitches in an instrument known as the Jew's harp, and in Italy, the *scacciapensieri*, which can be translated as "chases away worries." As a player changes the shape of her mouth, she changes the frequency resonance and accentuates different harmonics. It's intuitive to play, as we are used to changing these resonances in order to form the sounds of different vowels.

Listening #3

The repertoire for virtually every string instrument includes the use of harmonics. The most extreme example is a northern Thai (Lanna) instrument, the *pin pia*, made with a bow attached to half a coconut shell that is cupped and uncupped from the performer's breast to change the resonant overtones, like a guitar wah-wah pedal. For the pin pia, the only pitches played are the fundamental and its harmonics. There are no commercial recordings I am aware of, but two pieces by the virtuoso Cun Smithitham can be heard on my website (http://davesoldier.com/field.html).

On the classical harp, the first harmonic is typically the only one played, although it is certainly adaptable to the smaller divisions. The jazz violinist Stéphane Grappelli incorporates the third and fourth harmonics into his melodic improvisations, producing a sound of cascading waterfalls. Grappelli did not play anything the same way twice, but he often featured harmonics on a signature number, "Daphne."

Niccolo Paganini advanced a technique to play the fourth harmonic at any note on the violin. Listen to Paganini's *Variations on Rossini's Moses in Egypt* or the third movement of his First Violin Concerto in D Major, opus 6, for amazing harmonics.

A stunning use of harmonics was developed by the jazz and country and western guitarist Lenny Breau, who played dizzying melodies that alternated between fundamental and harmonic notes. Try "Mercy Mercy Mercy."

Listen to the shofar, which is made from a horn and used in Jewish religious prayer. The Yemeni services use a horn from an African antelope, which has a particularly beautiful tone.

Check out any Steve Turre piece in which he plays the conch shell. A nice one is "Sea Shell," with Shuichi Hidano and his taiko drummers.

Experience—maybe just briefly—commercial plastic "vuvuzela" horns played by the fans during the 2010 World Cup in Capetown, South Africa.

A fantastic example of using many horns of different sizes with natural harmonics to produce complex music is from an ensemble of instruments known as *ongos* playing "Ndraje balendro." It can be heard on the UNESCO Folkways record *Banda Polyphony* recorded by Simha Aron. In this ensemble, each player plays second and third harmonics (I think) on horns made from tree roots.

An early exception to a early classical trumpet being limited to the harmonic series is Haydn's Concerto for Trumpet in E-flat (1796). Haydn wrote it for Anton Weidinger, who used keys that worked like a flute to allow the air to escape before the open end of the trumpet. The keyed trumpet was never widely adopted, and the concerto became a hit only after valves were added to brass instruments in the 1830s. Markus Wuersch plays the trumpet concerto in the original way, on a trumpet with keys rather than valves.

Mari Kimura is a virtuoso violinist who developed a way to play *undertones*, that is, harmonic wavelengths *longer* than the fundamental, on her low G string. It seems that the typical node at the bridge is transformed into a vibrating antinode by appropriate pressure and manipulation of the bow, essentially doubling the wavelength of the string, similar to the way the clarinet produces a node in the air in front of the instrument.

Some trumpeters, like Cat Anderson and Maynard Ferguson, specialized in the very high harmonics of the trumpet (possibly to the thirty-second? It's hard to make out notes that high and just as hard to imagine all thirty-three nodes inside the trumpet tube . . .). They developed thrilling repertoires based on moving toward and away from these very-high-frequency multiples. For Cat Anderson's high notes with the Duke Ellington Orchestra, try "El Gato," and for Maynard Ferguson and a genuine upside-your-face experience of the 1970s, his hit version of the theme song to the movie *Rocky*, "Gonna Fly Now."

For the opposite extreme, the trumpeter Arturo Sandoval uses the trumpet's lowest frequencies in his solos, which trumpeters call "pedal tones."

For extreme New Orleans trombone low-note growls, check out Sidney Bechet's "Tailgate Ramble" by the Preservation Hall Jazz Band. The Dirty Dozen Brass Band is a contemporary New Orleans brass band with a low-note trombonist. They have inspired further generations of "second line" bands: try the Lil Rascals ("Buck It Like a Horse"). Trombone Shorty is a newer star of the tradition.

Inspired by Jimi Hendrix, who used a "whammy bar" to tighten and loosen his guitar strings and so change the pitch (try his "Star Spangled Banner"), the flutist Robert Dick developed a "glissando head joint" to increase the length of the flute and thereby reach lower frequencies. Try "Sliding Life Blues."

In Sicily, the Jew's harp, or jaw harp, is called the *marranzano*. In southern India, Carnatic classical music is performed on a version of the Jew's harp called the *morsing*. Virtuosos include N. Sundar and Nadishana.

4

• • • •

The Math of Sound and Resonance

- Why do sounds sound different? Why do voices?
- Are there mathematical definitions of noise and consonance?
- If sound exists, does antisound?

Now that we have learned about waves in the air and in strings and a bit about harmonics and resonance, hold onto your hats. You now have the foundation to understand how waves produce the wide world of different sounds and the physics underlying noise and consonance.

A singer, a violin, a church bell, an oboe, and a clarinet can each play a long note at the same fundamental frequency, yet we can easily identify the origin of each sound. This is because the different flavors and qualities of sound are formed from the way that the harmonics, and as you will see, nonharmonic waves, add together.

Waves in a bathtub provide a good illustration of how waves interact. We can drop a rock in a bathtub and watch the "spreading" waves radiate in a circle. Now drop a pebble at two different points and watch the waves collide and then move past each other.

Calculating the peak and trough of colliding bathtub waves is simple: add them. If the wave amplitudes are both one centimeter at the point where they collide, the height at the collision will be 2 centimeters (this is known as *constructive interference*, which you can hear in the hallway experiment in chapter 3). If the troughs of the waves collide, the height will be −2 centimeters. Where a peak and trough of the same size collide, the height will be 0 centimeters—they cancel each other out (*destructive interference*).

You're already a veteran when it comes to adding up sine waves: you do it when you look at a clock to tell the time. "Wait, Mr. Know-it-all," you say, "that's ridiculous! You don't add clocks." But your (mechanical) watch has an hour hand that rotates from 1 to 12 hours, and minute and second hands that rotate from 1 to 60. At any point, we tell the time accurately to the hour, minute, and second.

This means that the precise time in hours, minutes, and seconds of a mechanical watch determines time by summing three harmonic sine waves:

12 hours: 2 (rotations/day) / 86,400 (seconds/day) = 0.00002 Hz.
1 hour: 1 (rotation/minute) / 3600 (seconds/minute) = 0.00028 Hz.
1 minute: 1 (rotation/minute) / 60 (seconds/minute) = 0.016667 Hz.

The minute hand has a period 12 times smaller than the hour hand, and thus is the 12th harmonic of the fundamental frequency f_1. The seconds hand is 720 times faster than the hour clock, and thus is the 720th harmonic.

The realization that *any* periodic wave can be formed by combining some quantity of fundamental sine and cosine waves and their harmonics is credited to the French mathematician Joseph Fourier (1768–1830). The calculation of how much of each harmonic wave is added together to compose a sound or any wave is called a Fourier transform.

SIDEBAR 4.1

Joseph Fourier was an advocate for the French Revolution who later supported Napoleon and accompanied him on an Egyptian expedition in 1798. They discovered the Rosetta Stone, and Fourier showed its inscriptions to the eleven-year-old Jean-Francois Champollion, who used it to decode hieroglyphs some twenty years later. Fourier also discovered the greenhouse effect, in which gases in the atmosphere are responsible for the earth being far warmer than predicted, given the planet's distance from the sun, and so identified the basis for global warming.

The Harmonic Spectrum

A string on a musical instrument has two nodes that are always present, the bridge and the nut, that stop its vibration. As we discussed, the standing wave of the string also vibrates with additional nodes and antinodes in smaller-number divisions. As listeners, we tend not to be aware of these and focus solely on the fundamental frequency. However, the amount of each harmonic mixed into the full wave is what provides the distinctive qualities of the instrument and creates the vast palette of sounds we hear. Every performer of a string or wind instrument spends years of practice on producing long tones, and while they don't think of it in this way, they are learning to control the *spectrum* or balance of these harmonic frequencies.

SIDEBAR 4.2

Isaac Newton (1643–1727) introduced the term *spectrum*, the Latin word for "apparition," as in "specter," to describe the wavelengths of visible light, which he viewed from sunlight through a prism.

He wrote: "For as Sound in a Bell or musical String, or other sounding Body, is nothing but a trembling Motion, and in the Air nothing but that Motion propagated from the Object, and in the Sensorium 'tis a Sense of that Motion under the form of sound; so Colours in the Object are nothing but a disposition to reflect this or that sort of rays more copiously than the rest."

Realizing the parallel with frequencies of music, in his *First Book of Opticks*, he writes that the colors in the prism, red, orange, yellow, green, blue, indigo, and violet, correspond to the seven spaces between the notes of an eight-note just-intonation musical scale with the frequencies 1, 9/8, 6/5, 4/3, 3/2, 5/3, 16/9, and 2/1.

The easiest way to observe how the sound of an instrument is formed from harmonics is to use a program that runs a fast Fourier transform on a sound, resolving it into sine and cosine wave components (free programs at the time of this writing include Spear, Raven Lite, and Audacity). Most musical instruments and human voices display the sort of spectrum you might guess from the harmonic series, with a lot of the energy at f_1 and prominent bands with less energy at precise small real number multiples.

The construction of the instrument controls how much of each harmonic is present.

Recall that, for the flute, the fundamental frequency f_1 corresponds to a wavelength twice the length of the tube. The flutist, by controlling her embouchure, can deliberately play a second harmonic, the octave, by adding a higher f_1 between the air particle node and antinode in the middle of the tube. This means that we now have two places in the tube that vibrate between nodes and antinodes, and a frequency of $2 \times f_1$, corresponding to a wavelength exactly the length of the tube. Some flutists have learned to continue to add additional vibrational node/antinodes within the tube to produce an extended range of harmonics, effectively using the flute like a French horn, and this technique is an important addition to the flute's modern repertoire.

But even without intentionally playing higher harmonics, some of the energy of the air wave in the flute creates two, three, and four nodes and so on, producing the entire harmonic series (to a point; generally the components of roughly 8 Hz and above contribute little). As in figure 4.1, a Fourier spectrum demonstrates that many fundamental frequency multiples are components of the flute sound. While we interpret f_1 as the "note" the flute plays, the presence of those harmonics makes up the sound that we identify as "flute."

In contrast, remember that the clarinet has one closed end—the mouthpiece is always a node, and the open end is always an antinode—and that f_1 is four times the length of the tube. The second harmonic would require a wave half as long as f_1, which is twice the length of the tube, but that *can't happen*, as it would require either a node at the open end of the clarinet or an antinode at the closed end (these rules are also known as *boundary conditions*). For this reason, in contrast to the flute, the clarinet produces no (or very little) second harmonic. The third harmonic, $3 \times f_1$, requires a wave one-third as long as the fundamental: this *can* form with a node and antinode at the two ends. For this reason, a clarinet should only produce odd number multiples of the fundamental (figure 4.1).

Saxophones have a more conical bore (remember the previous chapter), and saxophonists *can* produce a node in the middle of the instrument, resulting in an octave when blown harder. They can finger the scales in the higher octave in exactly the same way as the lower octave. But because clarinet players can't produce an even number octave by overblowing, and as the next odd number interval is $3 \times f_1$, $= f_3$, an octave and a fifth, they need to learn new fingerings for playing the notes in the next-higher octave.

If you add more and more odd number multiples of a frequency, you produce a *square wave*, which produces a buzzy sound and is responsible for the acidic timbres of clarinets. A square wave, like a sine wave, can be modeled by watch hands on a clock, but in this case, the hand would start at twelve o'clock and stay there for a while before jerking to 6 o'clock, where it would stay before jerking back to twelve.

Beating, Consonance, and Noise

Intervals of octaves and fifths are used throughout the world in musical scales and even by the prehistoric cave flutes mentioned in chapter 2. In classical Indian music, the octave and fifth are sounded throughout the entire piece as a drone on a tambura or sruti box. In most but not all con- temporary musical styles, a "triad" consisting of octaves, fifths, and either minor or major thirds are heard by the audience as consonant and used to end musical phrases with a sense of completeness.

Adding together octave sine waves produces the remarkable feature that every node of the fundamental frequency f_1 remains, while half of the peaks are reinforced, resulting in a repeating, "periodic" wave. When a fifth, 3/2, is added to a fundamental, two-thirds of the nodes remain. Adding 5/4 major thirds or 6/5 minor thirds continues to produce a peri- odic wave.

Thus, when harmonics are sounded together, the wave in the air repeats regularly. While consonance and dissonance are to some extent culturally conditioned and subjective, the small real number harmonics have been generally considered to be consonant for thousands of years (figure 4.2). (Panbanisha, a bonobo coached to play on a synthesizer keyboard by Susan Savage Rumbaugh and Peter Gabriel [chapter 11], would choose to play octave notes together, so it may be still far more ancient.)

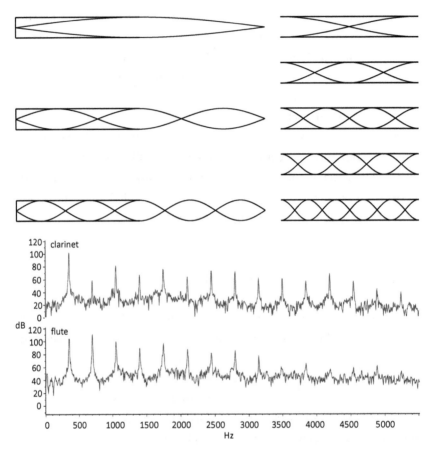

FIGURE 4.1 Flute and clarinet harmonics

As we saw in the previous chapter, for the clarinet modeled at upper left, the closed end is always a node, and the open end is always an antinode. A node cannot form in the middle of the tube because that would require the open end to be a node as well. A node can, however, form at two-thirds the length of the tube, thus dividing the fundamental wavelength by 3, producing f_3. Nodes also can form at two-fifths and four-fifths of the length of the tube to divide the wavelength in five, for f_5, and then at 2/7, 4/7, and 6/7 to produce f_7 and so on for odd-number nodes. The sound of a clarinet should therefore mostly be made of odd-numbered harmonics.

A flute, modeled at right, has two open ends and produces two high-vibration antinodes with a low-vibration node in the middle. Additional nodes can form, with two nodes that divide the flute to produce f_2, three nodes to produce f_3, four nodes for the second octave f_4, and so on. The sound of the flute should then be made of both odd and even harmonics.

How close are these models to reality? The upper trace shows the frequency components of a clarinet playing a fundamental f_1 of F4 (350 Hz), with the frequency in Hz displayed on the x-axis and the volume in units of dB on the y-axis. A small amount of even harmonics creeps in from the player's lips vibrating on the reed, providing a bit of open end at the mouthpiece. Still, consistent with theory, the f_1 peak is about +40 dB higher than f_2. Recall (chapter 1) that every 20 dB corresponds to a tenfold difference in amplitude, and so the sound wave amplitude of f_1 is $10 \times 10 = 100$ times higher than f_2. The f_3 and f_5 peaks are about 10 dB higher than f_4 and f_6.

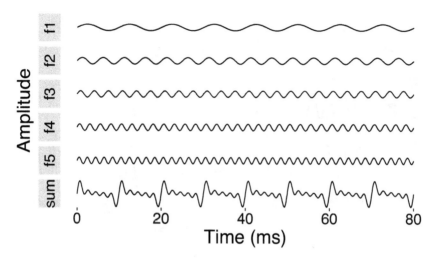

FIGURE 4.2 Harmonics add to form a periodic wave

A fundamental wave f_1 of 100 Hz and its next four harmonics (200 Hz, the first octave; 300 Hz, the fifth above; 400 Hz, for the second octave; 500 Hz, for a major third). The air pressure amplitude is shown on the y-axis and time in milliseconds on the x-axis. When these waves are added together, the resulting wave peaks are periodic and coincide every 10 msec. The composite wave sounds "in tune" and "consonant" and suggests the sound of a single note played on a musical instrument.

Source: Art by Jai Jeffryes.

In contrast, for the tritone, which you will remember is obtained by dividing the octave by the square root of 2 and is pretty universally considered to be dissonant, most periodic peaks and troughs no longer survive. Altogether, the scale intervals that are considered less consonant are also less periodic than harmonics based on whole-number ratios (see figure 4.3).

When the difference between even and odd frequencies in the first six components are added together (40 + 10 + 10 dB = 60 dB), the sound of this clarinet is made nearly exclusively of odd-numbered harmonics (60 dB = 10 × 10 × 10 = 1000-fold higher wave amplitude of odd harmonics), although odd- and even-numbered harmonics contribute about the same power at higher frequencies.

The lower trace shows the components of a classical flute playing the same fundamental as the clarinet, F4 (350 Hz). You can see the first nine harmonic peaks at multiples of 350 Hz. The first and second harmonics are about equally loud, and there is a gradual decrease in volume at higher harmonics, with little sound energy at peaks above 3.5 kHz. Together, this relatively similar contribution of odd- and even-numbered harmonics makes up the sound of this flute.

Source: Art by Jai Jeffryes. Used with permission.

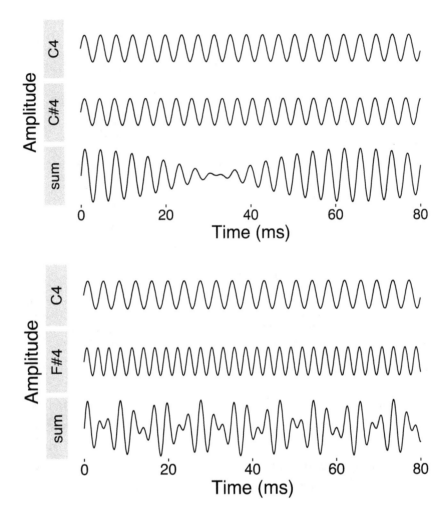

FIGURE 4.3 Waves at nonharmonic scale intervals

In the upper trace, the interval of a minor second, here an A4 440 Hz wave with the equal-tempered B♭4 above, produces beats where the waves destructively interfere. In the lower trace, a tritone ($f_1 \times \sqrt{2}$) interval of A4 440 and the E♭5 above is far less periodic than are harmonics. The resolution of this tritone "dissonance" to a consonance provides the harmonic basis of much of classical and contemporary music, including the "Great American Songbook," ranging from Stephen Foster to contemporary pop.

Source: Art by Jai Jeffryes. Used with permission.

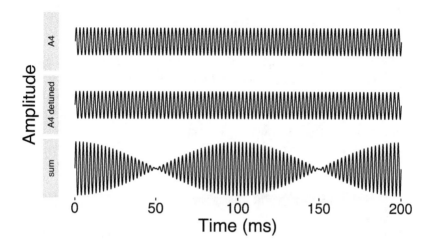

FIGURE 4.4 Beats arising from small pitch differences

A 440 and a 430 Hz sine wave played together. The combined wave peaks start in phase and constructively interfere to make the sound louder, but as the 440 wave moves ahead, the peaks and troughs overlap and destructively interfere to cancel each other out, so that the sound briefly disappears. This happens 440 − 430 = 10 times per second, and we hear 10 Hz beating.

Source: Art by Lisa Haney. Used with permission.

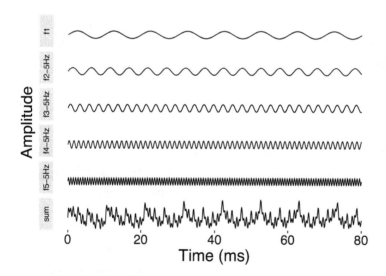

FIGURE 4.5 Out-of-tune harmonics

The sound wave resulting from adding f_1 at 440 Hz with f_2 through f_5, each detuned by 5 Hz. The resulting wave is mostly nonperiodic and sounds dissonant.

Source: Art by Jai Jeffryes.

SIDEBAR 4.3

Vincenzo Galileo (1520–1591) was a composer and lute player, and his son, Galileo Galilei (1564–1642), played the lute and keyboards. The younger Galileo addressed the mathematics of music in his final book, *Discourses and Mathematic Demonstrations Concerning Two New Sciences* (1638), written while the Roman Inquisition forbade him to publish new writing because of his discoveries about the solar system.

His explanation of the consonance and dissonance of musical intervals is similar to ours, and he illustrates the principles using diagrams of strings and descriptions of balls swinging on a pendulum.

You will remember the Pythagorean theorem and the story from chapter 2 in which Hippasus discovers, with deadly consequences, that the long side of a triangle can be a square root of two, an irrational number. This very number is what Galileo writes about in a discussion of the "harshness" of the tritone: "Especially harsh is the dissonance between notes with incommensurable frequencies; when one has two strings in unison and sounds one of them open together with a part of the other that bears the same ratio to its whole length as the side of a square to its diagonal; this yields a dissonance similar to the tritone."

Galileo was particularly fond of the real number "commensurable" vibrations of the perfect fifth, writing: "Thus, the effect of the fifth is to produce a tickling of the ear drum such that its softness is modified with sprightliness, providing at once the impression of a gentle kiss and a bite."

What happens when sound waves are out of tune? This is easiest to perceive when two pitches are a bit off from each other and produce audible *beats* as the waves alternate between peaks that superimpose (constructive interference) and troughs where destructive interference makes the sound briefly disappear (figure 4.4). From hearing beats, there is such a thing as *antisound*, and we "hear" it as the beats noticeable from the gaps in the wave. Musicians tune instruments by trying to abolish these antisound beats. If we continue to add more out-of-tune waves, meaning nonharmonic frequencies, the sound becomes still less periodic and is on the way to sounding like *noise* (figure 4.5).

The Colors of Noise

About a century and a half ago, the disturbingly brilliant Hermann von Helmholtz provided a definition of noise as *the perception of irregular motions of the air.*

By detuning harmonics, you hear that noise has a mathematical definition. It means that a sound is nonperiodic, which is equivalent to saying that the component waves are not the harmonics of a fundamental frequency.

But is noise to be avoided in music? An appropriate answer is often "hell, no." The amount of noise in sound can extend from very little, for example, when a flute or vibrating string plays a long tone, which are constructed almost completely by harmonics; up through church bells, which contain multiple harmonics of several unrelated fundamentals; to the noise of a subway train, which is composed of many unrelated frequencies.

Instruments with irregular shapes, including church bells and the Australian didgeridoo, vibrate at irregular resonances and thus have surprising frequencies that pop out at unusual multiples. However, the "noise" of these instruments is central to their appeal.

Some types of noise have color names.

Black noise is occasionally used as a euphemism for silence.

White noise, roughly the sound of a "shush" *shhhhhh* . . . and of radio static, has a distribution of random frequencies that are evenly distributed throughout the range of hearing. "White" is an appropriate term, as white light is a mixture of a broad range of light frequencies. As each higher octave encompasses a twofold-greater frequency range (see chapter 2), there is twice as much sound in each higher octave; that is, the probability of any frequency f occurring is the same as any other, so that a 100 Hz wave shows up as much as a 1000 Hz wave.

Pink noise is a "sweeter" sound, and some people use it to relax. The goal here is to spread the frequencies equivalently in each octave, so higher-frequency components are progressively decreased to correct for the crowding in white noise. The distribution changes with the reciprocal of the frequency, $1/f$: there is 10 times more sound energy at 100 Hz than 1000 Hz.

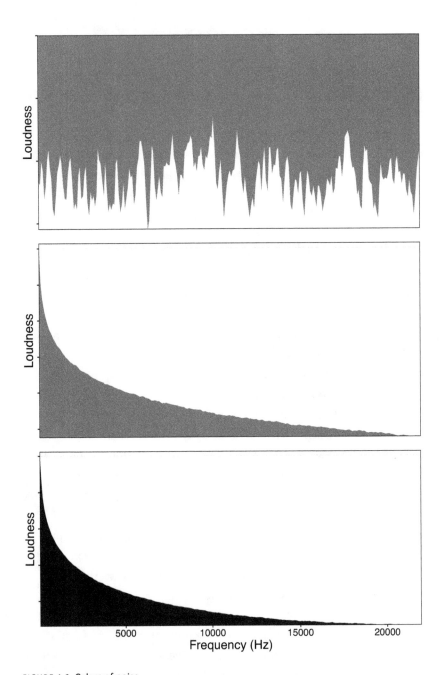

FIGURE 4.6 Colors of noise

A sample of white noise and the frequency distributions for pink and brown noise: frequencies from 20 to 20,000 Hz are displayed on the x-axis, and the relative loudness in dB is indicated on the y-axis. In contrast to strings and wind instruments, we don't observe harmonic

Brown noise resembles the sound of waterfalls and has been advertised as a sleeping aid. It is named after a pioneer of the microscope, Robert Brown, who described *Brownian motion* from the movement of pollen particles. You can observe Brownian motion in bright sunlight as the jerky movements of small dust particles in the air. Brownian motion has been used to describe everything from how a raccoon forages for food in the forest to the changes in value of an investment in the stock market.

In 1905, the same year he published the theory of relativity, Albert Einstein described Brownian movement as a "random walk," in which a particle jumps from one position to the next in a random direction. For sound, brown noise means that there are short jumps from one frequency to the next by small steps in either direction. The random jumps must wander a longer distance to enter into higher octaves, so the wave frequencies are mostly in the lower octaves, and their distribution is $1/f^2$. This means that the bass frequencies of brown noise are more exaggerated than in pink noise and drop off quickly: since $1/100^2 = 1/10,000$ while $1/1000^2 = 1/1,000,000$, there is 100 times more sound energy at 100 Hz than 1000 Hz (figure 4.6).

Just to make this auditory phenomena still more interesting, in the real world, the components of harmonics and noise can interact with each other, producing "nonlinearities" including *partials*, which are additional frequencies that are not integer multiples of f_1, and even *subharmonics*, which are below the fundamental frequency.

Consonance and Dissonance in Music

The attribution of fundamentals and harmonic frequencies that sound together—naturally known as "harmony"—as consonant or dissonant has a

---

peaks at real number multiples. In white noise (upper trace in white), the components are randomly and equivalently spread throughout the frequency range. In pink noise, the distribution follows the reciprocal of the frequency, $1/f$, so that there is much less sound at higher frequencies. For brown noise, the distribution of sound follows the reciprocal of the frequency squared, $1/f^2$, so that high frequencies drop off even more rapidly.

Source: Art by Jai Jeffryes.

physical basis: how far they deviate from real number multiples. But musical styles can "tolerate" dissonance differently, so a sizeable element of learned experience is involved in our experience of consonance and dissonance.

An example is the final resting harmonic intervals in some flamenco styles, which is a triad plus a minor second and seventh. To use this harmony at the end of a piece is considered dissonant and incorrect in nearly all other musical styles, but in some contemporary flamenco, ending the phrase with any other harmony clearly sounds "wrong."

Differing tastes for real number harmonics and noise are fundamental to the design of instruments. For example, although the West African marimba, the *gyil*, generally uses a just-intonation eight-note scale, the instrument makers add gourds that vibrate in nonharmonic intervals to add noise and produce a buzzier note.

Some contemporary hip-hop uses digital processing to intentionally add a great deal of noise to formerly consonant sounds, and many sounds in guitar-based rock styles are similarly altered. Some tenor saxophone players devote themselves to developing a powerful buzz in their sound, sometimes singing a bit out of tune with the note they are playing.

The use of noise as a building block for composition was made explicit at the turn of the twentieth century, in Luigi Russolo's manifesto *The Art of Noises*. A couple of recordings of his pieces of his specially designed noise machines performing with an orchestra have survived. I am happy to report that they sound as aggressive and disturbing as ever—no one we know would use them as sleeping aids!

The Voice as an Instrument

The most sophisticated instrument we have developed for titrating noise, consonance, and harmonic and nonperiodic waves is our voice.

The mechanics that provide this control begin with the vocal cords. These are located on the sides of the larynx and are composed of a muscle known as the *thyroarytenoid* or *vocalis*. The muscle is covered by a mucous layer that contains high levels of the protein collagen, which is in turn surrounded by a thin skinlike epithelium.

To vibrate the vocal cords, one pushes air from the lungs through the space between the cords, known as the *glottis*. The vocal cords control sound frequency, similar to how we can stretch a rubber band to change its frequency of vibration. If we want to produce the note A4, we certainly can't pluck a rubber band with our fingers 440 times a second, but we *can* tighten it at the precise point where a single pluck would cause it to vibrate 440 times a second.

SIDEBAR 4.4

The rubber band was developed as a virtuosic musical instrument, the Vibra-band, by Stan Wood. He cut out a large rubber band from sheets of latex used in dental surgery, and by blowing on it (rather than plucking it) he produced a sound reminiscent of the trumpeter Don Cherry.

As the vocal cord muscles are stretched, the frequency of their vibration increases as they alternately open and block the pathway for air being exhaled through the glottis. The opening and closing of the vibrating cords produces peaks and troughs in the air, similar to the siren we discussed in chapter 1, which pushes air through a rotating hole.

The pitch of the voice is controlled by the frequency of these air waves. The volume is controlled by how much air is expelled from the lungs to "explode" when the glottis opens.

To control the stretching of the vocal cords and change the vibrational frequency, we use another set of muscles, the *cricothyroids*, which lengthen and stretch to tighten the cords like a rubber band: you can feel the contraction of the cricothyroids as you sing at higher pitches. With very taut cords, some women can vibrate as fast as 2000 Hz, about three octaves above middle C and in rare cases even higher.

Still another set of muscles, the *thyroarytenoids*, relieve the tension and bunch up the cords so that they vibrate more slowly. Men can sometimes vibrate below 83 Hz (Johnny Cash's low note in "Folsom Prison Blues"):

you can feel these muscles tighten just behind the Adam's apple as you lower your pitch.

When working well, vocal cords produce something close to a sine wave at their fundamental frequency. The rest of the throat, mouth, and associated areas resonate at their characteristic frequencies, analogous to the nodes and antinodes of a string instrument's body. In this way, the identifiable sound of your voice comes from your unique personal anatomy.

MATH BOX 4.1

Why does breathing helium out of a balloon make you sound like Donald Duck? The frequency of the vocal cord vibration that produces the fundamental is unchanged, and the harmonics are at the same multiples. However, the resonance frequencies of the voice change in helium because now the throat and mouth are filled with a gas in which sound travels at 927 m/s, rather than at 343 m/s as in normal air. Remember that $f = c/\lambda$, and so the resonant frequencies within the mouth and throat are now more than an octave higher and thereby reinforce the voice's high harmonics.

We'll later discuss the double larynx of birds, the *syrinx*, which allows them to sing two sounds at once. But humans *can* produce two or three pitches simultaneously, particularly by "throat singing," as especially developed in the Russian province of Tuva, north of Mongolia. There are multiple styles of Tuvan throat singing that rely on different techniques, including whistled harmonics or by vibrating neighboring membranes in the larynx known as "vestibular folds," an approach also used by some death metal singers to make a menacing satanic growl. The Tuvans and some singers of Tibetan Buddhist chants use these "false vocal cords" to produce pitches an octave lower than those sung with the vocal cords.

Listening #4

The *gyil* is a marimba-like instrument played in Ghana and throughout West Africa for which an additional buzzy noise is central to its sound. The premier player in the United States is Valerie Naranjo.

Listen to Stan Wood and to his fellow musical instrument inventor Ken Butler perform on the Vibraband, perhaps the least expensive virtuoso musical instrument in the world.

The didgeridoo has been manufactured in Australia from hollowed-out eucalyptus trees for thousands of years. Focus on the unusual frequencies that arise, like a church bell, from the irregularities in the bore. A fine player and ambassador of the instrument is Lewis Burns.

For classic examples of the low and high notes of the human voice, try "Folsom Prison Blues," sung by Johnny Cash, has a low E (E2), about 83 Hz, the standard low note for a bass. Mozart wrote the D2, a full step below Cash's E, for the aria "O, wie will ich triumphieren" in *The Abduction of the Seraglio*. Mozart's "Queen of the Night" aria, properly "Hell's vengeance boils my heart" from *The Magic Flute*, written for his sister-in-law Josepha Hofer, reaches a high F (F6): 1396 Hz. One exemplary performance is by Edda Moser. For an entirely different rendition, check out the sublime Florence Foster Jenkins.

Occasional singers can reach even higher notes than the Queen of the Night, including Mariah Carey, who sang a G6 during her Super Bowl performance of "The Star-Spangled Banner" in 2002. The opera singer Audrey Luna reports that she can hit the C above that, C7, close to 2,000 Hz. At a superhuman extreme, the Australian singer Adam Lopez uses what he calls a "whistling" technique to reach more than an octave higher, D♭8 (4435 Hz), the high note of the D♭ piccolo. To keep our species in its place, consider that some bats produce notes at 200,000 Hz, more than five octaves higher than Adam can.

The most developed techniques for the human voice to sing more than one note at a time are by the Tuvan throat singers. These encompass several different traditions. Listen to examples of the technique known as *kargyraa*, which produces a low frequency, and to *khoomei* and *sygyt*. Huun-Huur-Tu is a group of great acoustic musicians and singers who specialize in this style and tour the West. A rock band that features throat singing is the Hu, a Mongolian heavy metal group. Some death metal vocalists also produce these sounds; you might listen to the band Suffocation.

Regarding using resonance as a device for music composition, listen to Alvin Lucier's "I Am Sitting in a Room," in which he records spoken instructions to perform the piece. The recording is played back and recorded a second time, and then played and recorded a third. As this process continues, we eventually hear the resonances of the room as musical pitches that replace his words.

As all sound can be constructed from harmonics, shouldn't we be able to model speech and other sounds pretty accurately with instruments that can create the appropriate harmonic series? Peter Ablinger found a way to make a piano perform a nearly comprehensible reading of "A Letter from Schoenberg."

A history of "noise" in music might begin with the oldest surviving illustrated book, *The Book of the Amduat*, written five thousand years ago, relating the nightly journey of Ra through the underworld to rise each morning as the sun. The book was rediscovered after Fourier and his colleagues found the Rosetta Stone and learned how to decipher hieroglyphs. The eighth hour of Ra's twelve-hour underworld trip specifies the musical sounds of the underworld, including the sounds of drumming and castanets; animal sounds such as the calls of birds of prey, cries of tomcats, and swarms of honeybees; and noises such as storm winds. I adapted this ancient sound score as an opera, *The Eighth Hour of Amduat*, featuring Marshall Allen, the leader of the Sun Ra Arkestra, as Ra the sun god.

Luigi Russolo (1885–1947), in his article "Art of Noises" from 1913, separated the use of noise in music into six categories. For example, category 2 is composed of whistling, huffing, and puffing. Luigi invented instruments to produce noise that he called *intonarumori*. "Corale" and "Serenata," two pieces for classical orchestra and intonarumori, were recorded by his brother, Antonio Russolo.

Perhaps the most extreme example of machine-based noise was by the Russian composer Arseny Avraamov (1886–1944), who created the *Symphony of Sirens*. Premiered in Baku in 1922, the music was performed by the cannons, sirens, and whistles of naval ships docked in the Caspian Sea and, on land, by bus and car horns, factory sirens, cannons, artillery, and machines, with a band and choir singing the "Marseillaise" and the "Internationale." The piece was conducted using pistols and semaphore flags suspended from a tower.

Long after Luigi Russolo, a noise music movement using abrasive non-harmonic and loud sounds as building blocks emerged. Notable examples include the "industrial" rock band Einstürzende Neubauten, the percussionist Z'EV, Lou Reed's *Metal Machine Music* (I am proud to have been fired as a college radio DJ for playing that record), Elliott Sharp, and Annie Gosfield. For Sharp, good examples are his Orchestra Carbon and his pieces for string quartet. For Gosfield, try *Burn Again with a Low Blue Flame* or *Rattling Beeps and Serging Sweeps*.

5

• • • •

Math and Rhythmic Structure

- How is math used to compose lyrics?
- How is syncopation defined?
- What were Bach's secret structural codes?

We've discussed the mathematics underlying Greek tunings for the lyre, but what of the math for lyrics? These calculations are called the *lyric meters*, the term arising naturally from the word *lyre* and from the Greek *metron*, "to measure."

Most current Western popular and classical forms, whether a waltz, hip-hop, march, stomp, stamp, or bourée, use meters made of repeating patterns of stressed pulses composed of only 2 and 3 counts. But many other styles, very much including those from ancient Greece, use *composite* or *compound* meters with alternating patterns composed from groups of 2s and 3s.

For Greek and Middle Eastern Lyric Rhythms, Use the Feet

The ancient Greek lyric meters were written in time divisions that the metricians called *feet*. Feet are formed of short syllables that last for one beat (*mora*), notated as ˇ (a *crochet*, ♩, or *quarter note*) and long syllables that last for two beats, written as — (a *minim*, ♩, or *half note*).

The iambic foot was one short and one long mora, (ˇ —). The infamous iambic pentameter, used extensively in English poetry (for example, most poetry by William Wordsworth), uses five iambic feet.

(ˇ)   — ˇ — ˇ   — ˇ   — ˇ   —
(♩)   ♩ ♩ ♩ ♩   ♩ ♩   ♩ ♩   ♩
(the) *chi*- ld *is* the *fa*- ther *to* the   *man*

We could either put more emphasis on the stressed syllables, which is known as *qualitative meter*, or, like the ancient Greeks, read with twice the duration on the stressed syllables, known as *quantitative meter*. If you try the second, you might easily compose a melody for this lyric in waltz time.

The Greek metricians cataloged a wide assortment of meters that extend far beyond those we use for contemporary lyrics. Their most popular was the dactyl foot, of one long and two short beats,

arranged in dactylic hexameter. That meter consisted of five dactyls,

$$— \breve{}\breve{} — \breve{}\breve{} — \breve{}\breve{} — \breve{}\breve{} — \breve{}\breve{} — X,$$

In which the X on the sixth foot could represent either a spondee,

$$— —,$$

or a trochee,

$$— \breve{}.$$

The dactylic hexameter was used for virtually all Greek epic poetry ranging from Homer to Hesiod in 700 BCE and for the following twelve centuries (!), including in the presumably sung lines of the *Iliad*, for which the melodies are lost. It is thought that performers dancing to this rhythm formed part of the show.

The dactylic hexameter is awkward in English, but you can almost sound it in this way:

For-ward the *light* bri-gade *charge* for the *guns* he said in-to the *Val-ley*

Why is it so hard to use dactylic hexameter in contemporary English? You can recite "The Charge of the Light Brigade" in quantitative meter, but to us, it is more natural to read it in qualitative meter with a count of three, like a waltz:

For-ward-the   *light*-bri-gade   *charge* for the   *guns*

Here the durations of each syllable are identical but a bit louder on the italicized syllables, again suggesting a waltz rhythm.

Ancient lyricists tended to be quantitative. Although we now tend to be qualitative, contemporary songwriters continue to use quantitative meter with longer durations on stressed syllables:

but the *fool* on the hill
La la *laaa* la-la *laaa*
La la *laaa* la la laa
La la *laaa* lala *laaa*

Meter becomes very challenging for us if we wish to compose in the styles of the superstar poet/singer/songwriter Sappho of Lesbos. We know that she sang her lyrics because portraits of her depict her singing with a lyre (figure 5.1). Sappho's song lyrics from about 600 BCE survive in more than three hundred stanzas, and they most often consist of three "hendecasyllabic" lines of eleven syllables of five feet (trochee, spondee, dactyl, trochee, spondee) followed by an Adonean foot, which was

FIGURE 5.1 Sappho

Sappho with an eight-string lyre and Alcaeus, another singing poet, holding a seven-string lyre, from a vase from 480–470 BCE attributed to the Brygos painter, redrawn on a flat surface by Valerie Woelfel.

Source: Valerie Woelfel. Used with permission.

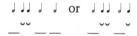

The standard rhythm for Sappho's lyrics is a killer for our poets and lyricists. Alfred Tennyson wanted to demonstrate that he had the skill to write English lyrics in Sappho's hendecasyllabic meter and came up with

— ˘   — —   —   ˘ ˘   — ˘   — ˘

♩ ♩   ♩ ♩   ♩   ♩ ♩   ♩ ♩   ♩   ♩

Ir-re-   *spon-si-*   *ble*   in-do-   *lent* re-   *view-ers*

Try your hand at the Sapphic challenge, but it's hard.

The Math of Composite Rhythms

Sappho's lyrics alternate between duplet and triplet phrases, producing a composite rhythm. While foreign to much of the West, composite meters are popular in contemporary Greece, the Indian subcontinent, southern Spain, and in the Middle East. Far from confusing audiences, these rhythms provide some of those regions' most popular dances and, even now, sung epic lyrics.

The epic lyric that is still widely sung in Greece, specifically in Crete, is the *Erotokritos*, written by Vitsentzos Kornaros in the 1600s. While it is "only" four centuries old, it provides insight into how older lyrics might have been sung by Homer, Hesiod, and Sappho.

The *Erotokritos* comprises ten thousand lines of fifteen syllables written in a dialect of eastern Crete. Only short sections are performed at a time. The most typical melody I have heard uses a composite rhythm of 2+2+2+3+2+2+4, providing a hypnotic groove of seventeen (!) beats. I am confident that the rhythm, which doesn't go stale when repeated many times, is a reason that this immensely long poem remains popular.

Composite rhythms are very popular in Middle Eastern and North African music. One of the best-known songs in the world, and the best known that survives from the pre-1492 era of Arabic Andalusia in southern Spain, is "Lamma Bada Yatathanna," attributed to Ibn al-Khatib of Granada (1313–1374). The lyric is in a poetic form known as *muwashaha*, which, along with the similar *zajal*, introduced the verse–chorus structure that provided the basis for the songs of the troubadours and, by that route, is the ancestor of most songwriting in the West.

"Lamma Bada Yatathanna" (figure 5.2) is performed in a ten-beat compound rhythm known as *sama'I taquil*, constructed from trochee, spondee, and trochee feet.

Middle Eastern rhythms are today often taught on the *dumbek*, a hand drum that produces a low *dum* like a bass drum and a sharp *bek* that resembles a snare drum. The ten beats of sama'I taquil on the dumbek are played (with hyphens for quiet or silent beats).

Dum __ __ bek __ dum dum bek __ __

In contrast to Greek quantitative meter, the goal here is to arrange the most stressed syllables to coincide with the strong beats on the *DUM*s and the beks.

FIGURE 5.2 "Lamma Bada Yatathanna"
Source: Author.

Flamenco continues the tradition of compound rhythms in southern Spain, particularly for a wide variety of composite twelve-beat patterns. These patterns are known as *palos*, after a walking stick used to tap out the rhythms, and each has its own personality based on lyrics, harmonies, history, structure, and subject.

For example, the *seguiriya* palo is associated with blacksmiths, one of the few professions Gypsies were allowed to pursue in Spain. It is sung in a Phrygian mode, and the rhythm was originally hammered on an anvil: I assume the seguiriya inspired Verdi's "Anvil chorus," which was sung by Spanish Gypsies in *Il Trovatore*, even though he wrote the piece in 4/4 in a major key.

The seguiriya uses a cycle of twelve beats formed as a composite of 2+2+3+3+2. The rhythmic pattern is sometimes taught using the name for an Andalusian blood sausage.

un   dos   *tres*   mor-*ci*-lla   mor-*ci*-lla

The *buleria* is presently the most popular flamenco palo that uses a composite rhythm, and I have watched as teenagers in Seville sing and play complex subdivisions in a city park for hours without a break. The rhythm, nominally a composite of 3+3+2+2+2, can be divided in the many ways that it is possible to divide cycles of 12. The dancers and musicians speak of playing "in 3" or "in 6" depending on where the phrases begin and end.

There are many clapping rhythms (*palmas*) for the buleria. The most common is counted and clapped as shown in figure 5.3, with foot stamps underlined on 3, 7, 8, 10, and 12. The strongest stresses are typically on the third beat, where the harmonies change, and the tenth beat, where the melodic phrases often end. Melodic phrases are often started around the twelfth beat.

The art of buleria is to play, sing, and dance phrases that divide this long count while never losing the pulse and ending together at the right place. For example, performers will switch to rhythms of four groups of 3s

FIGURE 5.3 A clapping rhythm for the *buleria*
Source: Author.

or two groups of 6s, or enter at unusual beats, or repeat a lopsided phrase several times but with the final phrase ending on the tenth beat. The most exciting point during a performance is when a long phrase arises in new patterns and the dancers, musicians, and clappers (*palmeros*) realize where to climax at the end of the phrase. Usually this is on the tenth beat, but sometimes the group realizes there is a syncopation and ends the phrase together on the ninth or on the ninth-and-a-half beat. When this is done well, the performers and audience shout encouragement to one another.

The most numerous composite rhythms are from the *tala* (from the Sanskrit for "clapping") system in India, and the flamenco tradition was in part derived from Indian rhythms, given the migration of the Gypsies out of northern India. The northern Indian Hindustani and southern Carnatic musical traditions differ in the names of the rhythms, but like the Middle Eastern dumbek, the sounds produced by the tradition's drums are used to define specific *tals*. In contrast to the two dumbek sounds, the northern *tabla* produces seven sounds, described by spoken syllables. For example, the *rupak tala* is a cycle of 7 made of 3+2+2:

Tin Tin Na *Di* Na *Di* Na

The cycles of tala meters as codified by Indian academics can run from two to 128 beats! Still, the long rhythms are generally divided into groups in 2s and 3s.

In Hindustani classical music, long, lopsided composite patterns can be repeated three times, to end up correctly on the correct ending beat, a technique known as a *tihai*. In the *chakradhar* tihai, the three phrases of tihai are themselves repeated three times, and sometimes even those sets are repeated three times, dazzling the audience as the piece finally ends at a complex but logical concluding beat.

Medieval Talea

The motet form is virtually extinct, but motets make up most of the earliest surviving extended compositions in Europe. The structure offers another form of rhythmic complexity, one similar to composite rhythms. While European classical music shed this approach over the centuries, the motet evolved into polyphonic (multiple voices at once) styles of classical music, including the fugues by J. S. Bach, and its influence continues.

A medieval ideal for the motet was to compose layers of melodic and rhythmic lines that are independent but work together when performed simultaneously. The recreation of their performance entails some guess-work, but early motets are thought to have been sung, with instrumental parts introduced as a later development.

A means to compose a motet is first to define a series of notes from a Gregorian church plainsong (*cantus firmus*), now often called the *color*. The line in which these notes were sung was called the *tenor* ("to hold"), the origin of the term "tenor voice." To provide the rhythm for the tenor line, a composer would use a rhythmic pattern known as an *ordo* or *talea* (which indeed is related to the Indian tala) that repeats throughout the composition.

Figure 5.4, an example of mapping two patterns of color and ordo, is from a motet by Guillaume de Machaut (~1300–1377). The top line shows the rhythmic pattern of the ordo, the middle indicates the note sequence color, and the bottom how they combine to produce the tenor.

The composers presumably wrote or borrowed the tenor line first. The additional lines in these motets, the *duplum*, *triplum*, and sometimes *quadruplum*, were typically more freely written but remained in accordance with the rules for harmony of the period, when consonances were considered to be limited to octaves, perfect fourths, and perfect fifths.

A feature of some motets, including Machaut's "Hocket David" and an anonymous motet known as "In seculum," is that the motetus and duplum have rests that break the flow of the melodic line, but when played together, a melody emerges that jumps between the voices. This style is known as a *hocket*, and some think the word has the same root as "hiccup." The hocketing of melodic lines is now unusual in most contemporary classical and popular styles but is a major structural element of vocal and instrumental music in central Africa.

FIGURE 5.4 Derivation of a tenor line in a motet

The top line is an *ordo*, a sequence of note durations, consisting of twelve sounded notes and five rests, followed by the *color* line that specifies the order of notes, in this case eighteen notes. Guillaume de Machaut mapped the color twice and the ordo three times so that they end together. This mechanism produced the tenor line of the motet "De bon espoir–Puisque la douce–Speravi."

Syncopation

The construction of a motet in which two different patterns are overlaid naturally leads us to consider the overlap of two different rhythms. This result is syncopation, a feature of rhythm ranging from the backbeat of most of contemporary Western popular music to the extraordinary complexity of the religious music of West Africa. These simultaneous rhythmic patterns typically result in a strong beat of one pattern falling on a weaker beat of the other.

A simple syncopation can be constructed by "phasing," in which two identical rhythms are shifted apart. A familiar example is when a bass drum plays a steady beat and handclaps or a tambourine play halfway between the drum's beats: this is the well-known "offbeat." If the handclaps are delayed a bit from the precise halfway point, we "swing" the beat: the offbeat shifted roughly two-thirds of the distance after the basic beat is a familiar sound in jazz drumming and melody.

Another common syncopation is to play two even beats in one melodic line and three even beats in another. This pattern, known as a *hemiola*, generally takes a while for pianists and drummers to learn, but it provides the rhythm for many Chopin compositions and is an important part of the vocabulary for jazz pianists such as McCoy Tyner.

Outstanding studies of syncopation were conducted by a missionary, Reverend A. M. (Arthur Morris) Jones, who studied Yoruba and Ewe drumming in West Africa and transcribed very complex syncopations. Jones called these *cross-rhythms*. To determine how the individual lines combined, he asked the musicians to perform each part individually. As with hockets, the different voices combine to form an overall, more sophisticated pattern. Also reminiscent of motets, the repeated drum phrases are often different lengths: for example, a *kadodo* bell performs a complex version of the hemiola, in which a compound pattern of eight eighth notes are played while other drums play patterns of three eighth notes, so that the phrases together require $3 \times 8 = 24$ beats to begin together again.

The drumming in these and other West African religious chants provide the origin of the highly syncopated traditions of Cuban, Haitian, New Orleans, and "funk" drumming and, developing out of them, styles of contemporary dance music throughout the world.

In particular, the patterns played on Yoruba *agogo* bells are the antecedent of the *clave* (Spanish for *key*) patterns played in Cuban and salsa music. In these styles, once introduced, a clave pattern typically lasts throughout the entire piece no matter what else happens around it, and the musicians refer to it to determine where the beats should be stressed. There is a range of clave patterns. For example, in the New York salsa style, the clave of a piece in 4/4 can start on the downbeat, which is called "three two" and is identical to the "Bo Diddley beat," or it can start on the offbeat of the first beat, which is called "two three."

The agogo bell patterns in the Yoruba tradition often use seven rather than five strikes; these are called "standard patterns" by scholars of West African music. Most drummers in these ensembles have specific highly syncopated parts, and a master drummer improvises over the bed of interlocking rhythms to keep things still more interesting. The singing often features solo and choir parts with even more patterns superimposed. Beyond Africa, these approaches are performed in religious ceremonies in Cuba, Haiti, and Belize. A single piece can continue for hours with a world of variety generated out of the interlocking parts, as in rumba ceremonies in Cuba praising the saints.

The theoretician Joszef Schillinger, among other things George Gershwin's composition coach, attempted to develop a system to analyze syncopation. He mapped out syncopations on graph paper from multiple rhythmic patterns that eventually line up and repeat again, that is, "periodicities."

With this sort of approach, the amount of syncopation or cross-rhythm can be assigned a quantity from the number of beats required to complete a pattern. For instance, an offbeat could be written as two beats in one voice during the period of a single beat in another, or 2:1, while the classic swing backbeat would have a value of 3:1. The hemiola can be characterized as 3:2 and the kadodo drum pattern as 3:8. The product of these could be used to assign syncopation values of 2, 3, 6, and 24, indicating the number of time divisions before the patterns coincide. This approach is similar to how Fourier analysis describes the contribution of harmonics to sound.

This approach, however, can be a challenge: the New York clave pattern of five strikes within a four-beat measure has one note that falls on a sixteenth of a beat: should that be 16:4, for a syncopation value of 64? If the same clave is played against an instrument playing triplets, is that 16:12— or a daunting value of 192?

Comparing syncopations becomes confusing when portions of cycles become longer, as with the drummer Clyde Stubblefield in James Brown's group, who used phrases that could be quite long and occur at the end of four- or eight- or sixteen-bar phrases. This analysis is harder still for a full West African drumming ensemble incorporating many parts of different lengths that might each begin at different points. It seems virtually impossible to contemplate in a performance by Elvin Jones, the drummer in John Coltrane's quartet, where the patterns undergo constant shift. But artists don't fear a challenge, right?

Canons and Riddle Canons

The most extreme tradition of using math to create a large work from a small kernel is the *canon*, from the Greek *kanon*, meaning "rules" or "law." In canons, a rule modifies an original phrase.

The most familiar rule for a canon is the *round*, also known as a *perpetual canon*. One voice plays a phrase (the *leader*), and while still playing, another voice or voices play the same phrase (the *followers*). The entire piece can repeat forever, or at least until the singers collapse in exhaustion. "Frère Jacques," "Three Blind Mice" (published in 1609 as a three-part canon in a minor key), "Row Row Row Your Boat," and "Hey Ho Nobody Home," must be the most widely sung polyphonic pieces.

The most familiar canon performed in concerts and soundtracks, Johann Pachelbel's (1653–1706, organist, composer, and teacher to the Bach family) Canon in D, follows similar rules but doesn't repeat *ad libitum*. He wrote the piece for three violins, with one leader and two followers, and adds greater elaboration of the original motif as it progresses. Pachelbel also wrote a bass line that is not repeated by other instruments, for a total of four voices. The resulting harmony from these four lines produces one of the most used harmonic patterns in current pop music, appearing in perhaps hundreds of songs since the original again became widely popular in the 1970s.

There are many ways that parts can follow the leader in addition to simply repeating from a distance. For example, a follower can repeat the leader backward, so that the last note of the leader is played first (*retrograde*); or the follower might be what would be read if one turned the page upside down (*inversion*); or the follower might move to a higher note as the leader moves to a lower note, as if reflected in a mirror (*contrary motion*); or one might multiply (*augmentation*) or divide (*diminution*) the duration of the notes to make the repeats longer or shorter. These transformations were highly elaborated in fugues composed for the keyboard, and in the twentieth century Arnold Schoenberg and others used these types of manipulations to transform short themes and rhythms in their compositions.

Guillaume Machaut composed a three-part canon motet in the 1300s, "Ma fin est mon commencement et mon commencement est ma fin" (My end is my beginning, and my beginning is my end). In the tenor, the leader runs to the middle of the piece, and then the original sequence of notes is repeated backward so that the first note also becomes the last: this type of a retrograde manipulation is sometimes called a *crab canon*. When the duplum and triplum lines reach their halfway points, they repeat the retrograde version of each other, as in a mirror image.

A special category is the *riddle* canon, also known as the *puzzle* or *enigmatic* canon, intended for a special audience of composers to communicate with one another over the centuries. Here, only the leader is notated, along with clues for how the riddle can be solved to create the full piece: one must be a detective and figure out how this should be done.

For example, one clue is that if a clef and key signature is drawn upside down, this indicates that the music is to be read in the same order and that a melodic sequence written to descend in the leader ought to ascend in the follower, a device known as a *mirror canon*.

Another clue is when a sort of squiggle is placed at the end of the leader, which traditionally means that the repeat should start a step higher in the scale.

Another approach is to use more than one clef for the leader. In this way, a note in the middle line of the stave can read as a B in the treble clef, a C in an alto clef, and a D in the bass clef, which can lead to dense parallel slabs of sound. This approach, I have read, was used by Ornette Coleman in his giant orchestral piece *Skies of America*, which is said to be written as a single melody played by each instrument without transposition.

The all-time master of riddle canons is the all-time master of everything, J. S. Bach. The most celebrated are the ten riddle canons in *The Musical Offering*, written using a theme given to Bach during a visit by Frederick the Great of Prussia.

In the first riddle canon, the leader is written in the alto clef, indicating that C is the first line and that the key is C minor. At the end is the same clef and key signature facing backward. The answer to the riddle is to play the piece in the usual direction and simultaneously backward (retrograde), an example of a crab canon structure centuries after Machaut (figure 5.5).

Other riddle canons in Bach's collection are rounds or use followers formed from contrary motion. One climbs whole steps six times until the

FIGURE 5.5 The first riddle canon

From J. S. Bach's original publication of *A Musical Offering*. The title at top, "Canones diversi super Theme Regium," means "different canons on a royal theme." The label "Canon Ia2" indicates that this first canon is for two voices.

Source: J. S. Bach, *A Musical Offering*, 1747.

piece ends in the original key: recall from chapter 2 our discussion of how the interval of a major second repeated six times, a "whole tone" scale, produces the Pythagorean comma.

Canons may be ancient, but new rules continue to be devised. Steve Reich's "Clapping Music" is a canon that follows a simple rule: the leader is a twelve-beat African drum rhythm clapped in unison by two performers. After eight repeats, one player pauses for an eighth note rest before starting again, and this eighth-rest delay pattern continues until, by advancing twelve times, the performers return to the original unison. It's a challenge to perform, but that listeners can understand its design is one reason it works. (Another reason is that, as with flamenco *palmas*, the performers on stage intently go through a procedure consisting of nothing but structured clapping, but when the piece ends, the audience responds with *un*structured clapping, a powerful effect.)

At an opposite extreme, in which canons are used to produce music of extraordinary complexity, are the player piano studies by Conlon Nancarrow. Nancarrow wrote canons of such difficulty that they were unplayable by human musicians. He punched holes in rolls so that they could be performed by player pianos at his house in Mexico City. For example, his "Study 21, Canon X" has one line that begins at a speed of about four notes per second and another at thirty-nine notes per second. The faster line slows down while the slower speeds up. By the end of the piece, the tempi of the lines have traded, and the original slow line is playing a surreal 120 notes per second.

His "Study 33" uses tempo changes in the follower that are multiplied by an irrational number based on the square root of 2 (remember our discussion of how this drove Pythagoreans to distraction and possibly murder in chapter 2?).

The mathematical transformation of a leader in a canon has been limited to addition/subtraction, multiplication/division, and mirroring, which makes sense because this approach was developed before calculus was introduced by Leibniz and Newton around 1700. It's not hard, however, to transform a leader as a *derivative*, which measures the distance between the frequency of each successive note. This abolishes changes in melodic direction over long phrases and decreases the overall range of melody. In contrast, if one makes an *integral* of the leader, in which the distances are added, the frequencies quickly extend above our range of hearing.

The concept of *fractals* provides another new approach for canonic augmentation and diminution. Fractals were named and described by the mathematician Benoit Mandelbrot (1924–2010) as geometric patterns in which the same patterns are observed when an object is viewed at different scales. A classic example is a coastline, where the same amount of jaggedness can be seen when viewed from a vantage of miles or centimeters. Mandelbrot's book *The Fractal Geometry of Nature* shows how fractal patterns can resemble trees, the flow of blood in the body, and landscapes. This approach has been so successful that fractal transformations are used to create the landscapes of other worlds in science-fiction films.

The best-known simple fractal is the Koch snowflake (figure 5.6), in which the sides of an equilateral triangle undergo repetitive division to produce another equilateral triangle on each side, first producing a Star of David and then snowflake patterns.

To some extent, fractal transformations occur whenever a pattern is compressed or elongated. But for a deliberate application of a fractal approach to a leader, the followers can follow the same patterns of changing notes at different time scales. If this is repeated multiple times, a strange phenomenon occurs, as the music sounds similar when it is performed at different tempos, and the dimensions of musical time have a different, less clear meaning.

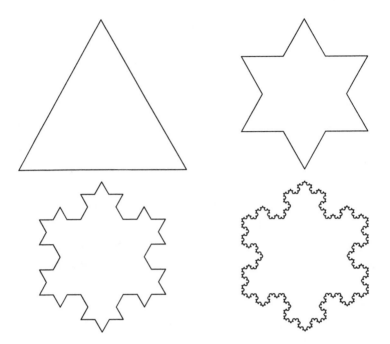

FIGURE 5.6 Koch snowflakes
The first stage is an equilateral triangle. The second stage adds another equilateral triangle on each side to produce a Star of David. Further iterations continue to add the same triangle to every side. The third and fifth stages of the Koch snowflake are shown in the bottom row.

Indeed, as explored by Mandelbrot and Felix Hausdorff, something happens in the math that describes geometric fractals such as the Koch snowflake, in which they can be formally proven to exist in partial dimensions, even when drawn on flat paper. With this bit of weirdness, in which the very dimensions of art and music are bent, one might glimpse possibilities for the future.

Listening #5

A lovely attempt at recreating all of the surviving ancient Athenian Greek scores is by the Spanish monk Gregorio Paniagua and his Atrium Musicae de Madrid, on the album *Musique de lal Grèce Antique*.

Here are some recordings to explore compound meters:

The Arabic Andalusian song "Lamma Bada Yatathanna," often credited to Lisan al-Din ibn al-Khatb from Granada in the 1300s, is an exemplar of composite rhythms as well as of the muwashahat poetic form, which is the ancestor of the verse–chorus structure of most Western popular music. This piece has been recorded hundreds of times. A particularly outstanding singer with fantastic musicians in this style is Fairouz, who is essentially the national singer of Lebanon, with the Rahbani Brothers.

The flamenco seguiriya, while in principle a composite cycle of 2+2+3+3+2, can be challenging to follow in real life given its tempo changes and the freedom of expression consistent with its mournful subject matter. Classic singers of the seguiriya include Manuel Agujetas and Manuel Torre.

All flamenco performers play bulerias, and any video of an Andalusian Gypsy wedding or *juerga* (flamenco fiesta) will demonstrate how it is performed as a series of improvised solos. The rhythm is nominally 3+3+2+2+2, but that pattern is then subdivided and syncopated in myriad ways. The possible ways to combine and subdivide the counts may be clearest to hear with guitarists, including Paco de Lucia, who was the preeminent innovator and master technician of the contemporary style, as well as Tomatito and Moraito, masters of syncopation within the rhythmic cycle.

A remarkable music film on Gypsy (or Romany) musical styles, with stunning performances and nearly no dialogue, is Tony Gatlif's *Latcho Drom*. It follows their migrations, the changes in their customs, and the evolution of the compound rhythms in their music and dance as they spread from the Thar Desert in northern India through Egypt, Turkey, Romania, Hungary, Slovakia, France, and Spain.

Of the incredible variety of Indian rhythmic cycles, some of the wildest talas have been released under Ravi Shankar's name and feature his percussionist Alla Rahka on tabla. These include the fourteen-beat 3+4+3+4 tala "Farodast," on the album *Homage to Mahatma Gandhi*.

The Dagar Brothers are members of a famous family expert in an ancient style known as *dhrupad*. Some of their pieces are in a style known as *dhamar*, which is associated with the Holi spring festival and played in a fourteen-beat 5+2+3+4 tala.

For American popular music, the best-known unusual compound meter hit is "Take Five" by the Dave Brubeck Quartet, which is five-and-a-half minutes of 3+2, with wonderful Paul Desmond saxophone and Joe Morello drum solos—yes, this was a hit record in 1959! It is said to be the best-selling jazz single of all time. The B-side, "Blue Rondo à la Turk," uses a Turkish compound rhythm transformation of a Mozart rondo. All of the pieces on the album *Time Out*, from which this single comes, use unusual rhythms.

For a pinnacle of syncopation, listen to music from the Ewe in West Africa. The classic recordings from A. M. Jones and others are great, and they demonstrate contemporary styles of choral religious music. The Ewe rhythms have been taught to a wider audience in the United States by the master Ewe drummer C. K. Ladzekpo, who directed the African music program at the University of California at Berkeley. He has produced some outstanding videos to teach these patterns.

To hear how the Yoruba and Ewe styles were transformed in New Orleans, you might listen to Ziggy Modeliste, of the Meters, and the drummers for the contemporary brass bands. Start with the Dirty Dozen. Another extension is by the drummers in James Brown's band, Clyde Stubblefield and Jabo Starks, around the period of *In the Jungle Groove*.

John Coltrane's *My Favorite Things* strongly features the two-against-three hemiola: listen to the pianist McCoy Tyner's left hand during his improvised solo. He uses the hemiola in alternate measures through most of the piece (the full rhythm under Coltrane's solos is usually four beats over a count of six; try to figure it out). Yes, this was a hit record in 1961!

I would guess that the most virtuosic of complex syncopation may be the playing of the drummer Elvin Jones during his membership in John Coltrane's quartet. He plays rhythms that can be counted in several ways; this is especially evident on live recordings such as "India," on the album *Impressions*, which was likely influenced by Hindustani tabla drumming. In some ways, this approach was taken even further by the drummers Sunny Murray and Dennis Charles, who played with Cecil Taylor, and the brilliant drummer William Hooker.

Chopin uses quite a bit of hemiola, for example in his Fantasie Impromptu in C-sharp Minor, a challenge to perform accurately, and it can be hard to discern the left-hand rhythms in three against the right-hand figures in four.

Gyorgi Ligeti's *Piano Etudes* feature multiple extrapolations of various hemiola approaches, with inspiration from West Africa, East Asia, Steve Reich, and Conlon Nancarrow.

To explore the isorhythms of the motet, listen to recordings of Guillaume de Machaut's hocket motet "Hocket David."

Here are some recordings to explore mathematical transformations of music, as exemplified by the canon:

Steve Reich's "Clapping Music" is a canon that follows an extremely simple and effective rule that is almost a round.

Brahms wrote a gorgeous canon for four sopranos, "Mir Lächelt Kein Frühling," in which the keys modulate in so subtle a way that most listeners do not notice it.

Timothy Dwight Edwards maintains a website on "puzzle canons" by composers throughout the ages, including ones by Albert Roussel, William Byrd, and John Dowland.

As mentioned, the most stunningly complex canons are by Conlon Nancarrow. Of the player piano etudes, I am particularly fond of "Study 12," which uses the flamenco Phrygian scale and somewhat imitates the guitar. "Study 21" is deservedly popular. "Canon for Ursula" is written for a human pianist, Ursula Oppens, to perform. I adore a chamber group version of some of these pieces, orchestrated by Yvar Mikhashoff and recorded by the Ensemble Modern.

A popular approach for a canon enabled by digital technology is simply to augment music by slowing it down. The software engineer Paul Nasca developed software to slow down recordings by running a Fourier analysis to smooth over the gaps in the waves and maintain the original pitches. Perhaps the most popular such canon is a manipulation of the Justin Bieber song "U Smile," by Shamantis, who slowed it to an eighth of its original speed. "9 Beet Stretch" is Beethoven's Ninth Symphony slowed to twenty-four hours by Leif Inge, using software developed by Bill Schottstaedt.

While most music is moderately "fractal," in that it uses repeating motifs and rhythms with phrases repeated at different pitch frequencies or with altered timing, one can deliberately pursue the goal of creating fractal music. The composer Adam Neely developed a clever and deliberate way to convert musical phrases to fractals, in a way that will be apparent to the listener during playback. He chose a short phrase from a song and repeated it at a very fast speed so that the repeats themselves produce a pitch. If a melodic phrase is repeated at 440 Hz, it will produce the note A4 (440 Hz), and if repeated at 220 Hz, it will produce the A3, an octave below. In this way, a melodic phrase can be heard, and if the tempo is slowed during playback the phrase disappears, but when played back at a still slower speed, the phrase reappears.

I wrote a literal fractal canon made of only the first five notes of the second movement of Arnold Schoenberg's second quartet for my String Quartet no. 3, *The Essential*. The movement "Benoit Meets Arnold in 5 Dimensions" consists of the five-note intervals at four speeds, with the pitch intervals superimposed on one another like a Koch snowflake. The notes follow the pattern precisely but don't repeat, and like a *tihai*, they catch up at the end of the pattern, which in the recording by the PubliQuartet takes about two minutes. Other canonic transformations in my Quartet no. 3 include a first derivative, a Fourier transform, and the integral, in which the pitches rapidly exceed the range of hearing.

Two freer fractal transformations, as the number of notes otherwise become dense quickly, appear in my collection of piano pieces *Fractals on the Names of Bach and Haydn*. This is in the tradition of using someone's name to spell a melody (Bach is spelled in notes as B♭-A-C-B) as the leader. My church organ version of the fractal piece that uses Haydn's name is "*Olivia porphyria*," a sea snail with fractal patterns on its shell.

Fractals also arise from the playback of the same sounds at different speeds. This can produce new musical features absent in the original material. My podcast for WFMU radio's People Like Us's (Vicki Bennett's) series *Timeless Music* goes through this with a bit of Frank Sinatra vocals.

I wrote *Variations on Chopin's "Minute Waltz"* based on an old joke about a bad piano player: "It takes him half an hour to play the 'Minute Waltz.'" The idea is to use newer math to produce canonic variations. The variations include (1) the *average pitch* of the notes: if there is a C and a C♯, we play a C a quarter tone sharp; (2) playing *every pitch except* the notes in the original score; (3) *no rhythm*, with every note played simultaneously; (4) a *Fourier transformation*, where the amount of each pitch in the piece is played without regard to rhythm; (5) *no pitch information* (the rhythm is played on one note); (6) the piece compressed into six seconds; (7) the *derivative*, where the distance between successive pitches is played; (8) the *integral*, in which each pitch is added to the previous one, so that the notes rapidly rise above the range of human hearing; and (8) the original score of the "Minute Waltz" played over half an hour, recorded live in concert, with the notes time stretched by Sean Haggerty (pity the audience at the premiere at Greenwich Village's Le Poisson Rouge concert hall).

CENTERPIECE *The Sense of Hearing*

Painted by Jan Breughel the Elder and Peter Paul Rubens in 1618 for a series on the senses. The woman may be Euterpe, the muse of music. The instruments include a two-manual harpsichord, drum, trumpet, trombone, cornett, a tenor cornett (known as a lizard), lute, shawm, flutes, viols, and gambas. On the table at right are musical clocks and small instruments and, underneath,

hunting horns. There are singing birds, a cockatoo, and two parrots. Pictured are several musical scores, including a four-part piece under a viol. One score is a six-part madrigal dedicated to Prince Pietro Philippi Albert and Princess Isabella.

Source: © Museo Nacional del Prado.

6

. . . .

Brain Mechanisms of Rhythm

- What are brainwaves?
- What creates our internal clock, and how does our sense of rhythm work?
- How does periodic electrical activity arise in the nervous system?

If we can sing and dance in rhythm, the nervous system must possess an internal clock. This clock needs to be able to respond to and adjust to the tempo of the music we are listening to. And for some styles of music and dance, we need to operate multiple clocks at once.

What are these internal clocks? Spoiler alert: there are big gaps in our knowledge. We'll discuss the basics, and if civilization continues, we will know more in the years to come.

Thinking About Current

The nervous system enables our senses and controls our behavior by activating and silencing a vast number of possible circuits. Its cells, known as *neurons*, make up the elements of the circuits and act like electrical batteries. The connections between the cells that activate or depress the cells' activity were named *synapses* by the British neurophysiologist Charles Sherrington (1857–1952).

SIDEBAR 6.1

You will remember from chapter 2 that the ancient Greeks called the note that doubles the fundamental to complete one octave and start the next the *synaphe*, which means *union*, while *synapse* means "unite." Charles Sherrington was also a published poet with an intimidating vocabulary and a 1925 collection, *The Assaying of Brabantius*. Was Sherrington aware of the ancient musical term?

MATH BOX 6.1

Batteries maintain a voltage difference across a resistor. Here's a metaphor that will help you think about batteries and neurons and get you acquainted with Ohm's law (described by Georg Ohm, 1789–1854), a simple mathematical relationship between current, voltage, and resistance.

Imagine a full water tank connected to an empty bathtub by a garden hose with a faucet. If the tank is placed at a higher elevation than the bathtub and the faucet is opened, water *current*, like that in a river, will flow from the water tank to the tub. The greater the difference in elevation between tub and tank, the greater the difference in water pressure. This is why water towers are built at a high elevation. (Each 10.2 cm increase in the height of a water tower increases the pressure by 1 kilopascal; that is, doubling the difference in water levels doubles the water pressure.)

The hose carrying water between the tank and tub is a *conductor*. The rate of the water current through the hose is measured in units of volume (1 milliliter = 1 cm³) per second. This is equivalent to the current of electrons or charged ions through a conductor, for example, a copper wire or a high-tension power line.

For water to flow continuously and without interruption through the hose, the flow rate through the intake must be the same as its efflux through the faucet. If the diameter of the hose is the same throughout, the velocity of the water moving through it (in cm/second) will likewise be the same throughout.

If we replace our garden hose with a hose that is twice as wide in area (measured in cm²), we will double the number of flowing water molecules;

(continued)

MATH BOX 6.1   (CONTINUED)

that is, we have increased the garden hose's *conductance*. Doubling the hose length (you might coil it like a Slinky) makes the transit of a water molecule from the water tank to the bathtub take twice as long.

Since the flow of water molecules is proportional to the molecules' velocity and the hose's area but inversely proportional to the hose's length, we can define conductance as

$$\text{hose conductance (cm}^2\text{/second)} = \text{water molecule velocity (cm/second)}$$
$$\times \text{ hose area (cm}^2) \text{ / hose length (cm)}.$$

The movement of water downhill is proportional to the difference in water levels, so the water current from the tank to the tub is

$$\text{current (cm}^3\text{/second)} = \text{difference in water level (cm)} \times \text{hose conductance}$$
$$\text{(cm}^2\text{/second)}.$$

Note that if either the water level difference or the hose conductance is increased, the current is faster.

If we substitute the terms electrical engineers use for current (I); potential energy difference, or *voltage* (V); and conductance (G), we obtain

$$I = V \times G.$$

Notice that as the voltage is lowered, the current decreases, as is intuitive from when a car battery begins to lose its charge.

The reciprocal of conductance is known as resistance (R).

$$R = 1/G.$$

Behold the standard version of Ohm's law, derived from a bathtub:

$$I = V/R, \text{ or, rearranged, } V = IR.$$

To make these calculations easier, engineers defined units of amps (amperes) for current, volts (naturally) for voltage, and ohms for resistance as

$$1 \text{ volt} = 1 \text{ amp} \times 1 \text{ ohm}.$$

One ampere of current is when one *coulomb* of electrons flows per second. A coulomb is a ridiculously large number: 6.241×10^{18} electrons (10^{18} is a billion billions). The flow of a billion billion electrons may be more difficult to imagine than the flow of a liter of water, but in neurons we will be dealing with much smaller quantities. Neuroelectrical currents are measured in nanoamps (a billionth of an amp and billions of ions) and picoamps (a trillionth of an amp and millions of ions).

The Nervous System Produces Electricity

If biology and electrical physics now seem like unrelated fields, they did not appear so to their pioneers. A central question for the early electrical scientists was whether animals produced electricity.

Johann Georg Sulzer (1720–1779) reported in 1752 that when his tongue touched two metals, lead and silver, together, he tasted something disagreeable that was not present when the metals were tasted separately. Sulzer's phenomenon is now called the "battery tongue test," and every electric guitarist who uses stomp boxes knows that you can perform it by touching your tongue to the two poles of a nine-volt battery. He conjectured that particles were being liberated from the metals that activated nerves in the tongue, which is true. I am happy to say that my namesake developed a way to detect electron flow by taste long before the invention of voltmeters.

The search was on for electricity that was produced, not simply conducted, by animals. A strong suspect was the torpedo, a cartilaginous fish closely related to the stingray and the namesake of the naval weapon. The torpedo fish was reported by the ancient Egyptians and Greeks to stun its prey and, occasionally, swimmers. The torpedo is also called the "electric fish," which sort of gives the answer away.

The physicist and polymath Henry Cavendish (1731–1810) was convinced that the swimmer's shock could be explained by the torpedo discharging electricity, and in a "stunning" paper published in 1776, he described how to build a model of the fish at home. In a variation of Benjamin Franklin's

1752 demonstration of atmospheric electricity by receiving a spark from a metal key on a wet kite string during a lightning storm, Cavendish demonstrated that his artificial torpedo could produce electricity by inviting visitors to touch the model and feel the shock (figure 6.1).

The quintessential demonstration of electricity in the nervous system was by Luigi Galvani (1737–1798), derived from experiments he began in 1780. He wrote in a 1792 publication for the Bologna Academy of Science that when his student touched a nerve of a dissected frog leg with a scalpel that had accumulated a static charge, it discharged a spark, and they saw the legs "contract so that they appeared to have fallen into violent tonic convulsions."

Galvani further adopted Franklin's approach by reporting that a frog leg jumped when connected to a lightning rod during a storm. The leg also twitched if he connected a dissected frog spinal cord with the attached leg to two metal wires made of bronze and silver. He reported that he had successfully repeated these experiments with the legs of dissected chicken and sheep.

Galvani called these phenomena *animal electricity,* and others later called it *galvanism* (in his honor) and *bioelectricity.* Whether triggered by lightning,

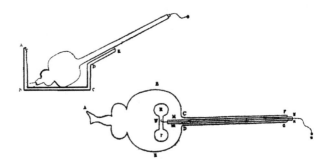

FIGURE 6.1 Cavendish's artificial torpedo fish

Henry Cavendish cut a piece of wood in the shape of a torpedo fish with a forty-inch handle covered in leather. A glass tube was placed between *M* and *N.* A wire at *W* was threaded through the tube and soldered to a strip of pewter intended to represent the electric organs. Below, this model is immersed in a bath of salt water into which Cavendish applied electricity from forty-nine Leyden jars, early capacitors that store and then discharge current. When an observer touches the wire, they feel a shock, and sparks would be seen to fly from the handle.

Source: Henry Cavendish, "An Account of Some Attempts to Imitate the Effects of the Torpedo by Electricity," *Philosophical Transactions* 66 (1776): 196–225.

touching two metals together in one's mouth, or a spark from a charged scalpel, Galvani believed that the twitches could also be explained by electricity created by the animal itself (figure 6.2).

SIDEBAR 6.2

Galvani's 1791 publication and its coverage in the newspapers—the *Morning Post* wrote a grisly story on Londoners repeating the experiment on a recently decapitated dog in 1803—inspired Mary Shelley's (1797–1851) *Frankenstein* (1818). She later wrote:

> Many and long were the conversations between Lord Byron and Shelley, to which I was a devout but nearly silent listener. During one of these, various philosophical doctrines were discussed, and among others the nature of the principle of life, and whether there was any probability of its ever being discovered and communicated.
>
> They talked of the experiments of Dr. Darwin (I speak not of what the doctor really did, or said that he did, but, as more to my purpose, of what was then spoken of as having been done by him), who preserved a piece of vermicelli in a glass case, till by some extraordinary means it began to move with voluntary motion. Not thus, after all, would life be given. Perhaps a corpse would be reanimated; galvanism had given token of such things; perhaps the component parts of a creature might be manufactured, brought together, and endued with vital warmth.

In Pavia, Italy, Alessandro Volta (1745–1827) reproduced Galvani's frog leg observation, but he claimed that frog legs were not required to produce the electrical current. To prove his point, he took disks made from different metals and connected them together with cardboard soaked in salt water, thereby inventing *voltaic piles*, now known as batteries. To demonstrate that this contraption produced electrical current, he used the Sulzer/Cavendish/Galvani/Franklin approach, reporting that one felt a shock when touching both ends.

It was thus proved that a biological organism was not necessary to produce electricity. From these results, by 1793, Volta denied the existence of animal electricity, precipitating a scientific skirmish that continued for decades.

FIGURE 6.2 Galvani's demonstration of animal electricity

Apparently influenced by Benjamin Franklin's kite experiment, Galvani used a lightning rod connected to a wire that caused the frog legs on the table to jump.

Source: Luigi Galvani, *De viribus electricatatis in motu musculari* (Bononia: Ex Typographia Instituti Scientiarium, 1791).

SIDEBAR 6.3

Volta's experiment was repeated in Britain by William Nicholson and Anthony Carlisle, who reported in 1800 that the voltage produced from Volta's battery could produce the gases oxygen and hydrogen from water. Volta and his colleagues thus launched the field of electrochemistry, which underpins virtually all contemporary electrical technology and the digital age.

Ultimately, electrical currents from nervous tissue were reported by the Italian physicist Carlo Matteucci (1811–1868), who, appropriately enough, used an instrument known as a galvanometer on the electric lobe of the torpedo electric fish. (Fittingly, the voltmeters and galvanometers named for Volta and Galvani are similar devices.)

FIGURE 1.4 The People's Choice: America's Most Wanted and Most Unwanted Paintings

The artists Vitaly Komar and Alex Melamid's most and least wanted paintings for the population of the United States, based on a 1993 poll of the preferences of 1001 American adults, sponsored by the Nation Institute. *The Most Wanted by Majority* was determined because 66 percent of the population preferred the color blue, 49 percent preferred outdoor scenes, 60 percent wanted paintings to be the size of a dishwasher (here, 24 × 32 inches), and 65 percent preferred inclusion of historical figures (hence George Washington). For *The Most Wanted by Minority*, sharp angle geometric forms, a small canvas (5 × 8.5 inches), and separated colors were unpopular. A similar poll for tastes in music was conducted in 1995 to produce "The Most Wanted Song and "The Most Unwanted Song."

Source: Photo by D. James Dee. Courtesy of Komar and Melamid and Ronald Feldman Gallery, New York. Copyright Vitaly Komar and Alexander Melamid.

FIGURE 2.3 Images of a begena and lyre

A muse playing a six-string lyre on Mount Helicon with a magpie at her feet (attributed to Achilles Painter, circa 334 BCE) and the contemporary Ethiopian musician Zegeye Asaye playing a six-string kraar.

Source: Vase image from Staatliche Museum Antikensammlungen, Munich. Zegeye Asaye photograph © 1995 Jack Vartoogian/Front Row Photos.

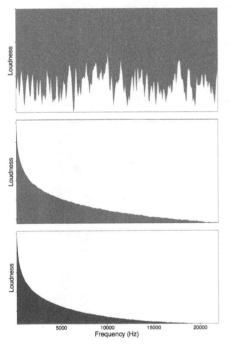

FIGURE 4.6 Colors of noise

A sample of white noise and the frequency distributions f pink and brown noise: frequencies from 20 to 20,000 Hz are d played on the x-axis, and the relative loudness in dB is indicat on the y-axis. In contrast to strings and wind instruments, don't observe harmonic peaks at real number multiples. In wh noise (upper trace in white), the components are randomly a equivalently spread throughout the frequency range. In pi noise, the distribution follows the reciprocal of the frequency, 1 so that there is much less sound at higher frequencies. For bro noise, the distribution of sound follows the reciprocal of the f quency squared, $1/f^2$, so that high frequencies drop off even m rapidly.

Source: Art by Jai Jeffryes.

FIGURE 5.4 Derivation of a tenor line in a motet

The top line is an *ordo*, a sequence of note du tions, consisting of twelve sounded notes and f rests, followed by the *color* line that specifies the or of notes, in this case eighteen notes. Guillaume Machaut mapped the color twice and the ordo th times so that they end together. This mechanism p duced the tenor line of the motet "De bon espc Puisque la douce–Speravi."

CENTERPIECE *The Sense of Hearing*

Painted by Jan Breughel the Elder and Peter Paul Rubens in 1618 for a series on the senses. The woman may be Euterpe, the muse of music. The instruments include a two-manual harpsichord, drum, trumpet, trombone, cornett, a tenor cornett (known as a lizard), lute, shawm, flutes, viols, and gambas. On the table at right are musical clocks and small instruments and, underneath, hunting horns. There are singing birds, a cockatoo, and two parrots. Pictured are several musical scores, including a four-part piece under a viol. One score is a six-part madrigal dedicated to Prince Pietro Philippi Albert and Princess Isabella.

Source: © Museo Nacional del Prado.

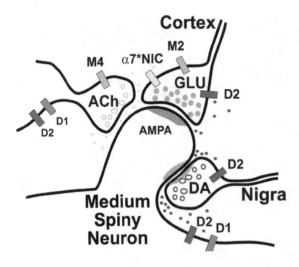

FIGURE 7.2 A striatal microcircuit

A illustration of a simplified striatal microcircuit indicates D1 and D2 dopamine (DA) receptors (these are on different popuations of medium spiny neurons); AMPA-type receptors for glutamate (GLU); and M2, M4, and alpha7 nicotinic receptors for acetylcholine (ACh). Even with only ten receptors indicated, this tiny circuit can exhibit $2^{10} = 1024$ states.

Source: Art by Nigel Bamford (Yale University). Used with permission.

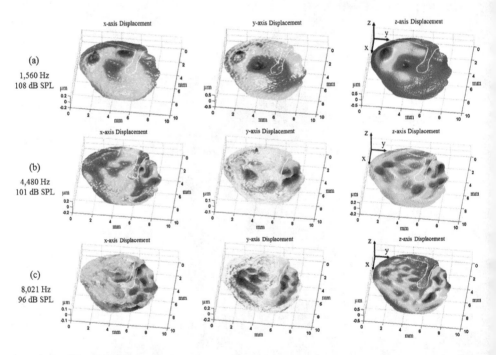

FIGURE 8.2 Vibration of the eardrum

In a contemporary version of von Békésy's approach, a human eardrum was stimulated at three different sound frequencies and the vibrations recorded by high-speed video. The rows represent three applied frequencies, and the three columns show the vibration in each spatial dimension, with red indicating the greatest and blue the least displacement of the eardrum membrane. The resonances resemble a snare drumhead.

Source: From Morteza Khalegi, Curlong Cosme, Mike Ravica, et al., "Three-Dimensional Vibrometry of the Human Eardrum with Stroboscopic Lensless Digital Holography," *Journal of Biomedical Optics* 20 (2015): 051028.

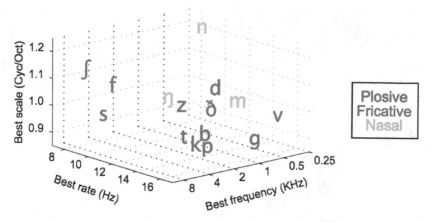

FIGURE 9.4 Neuronal response by the ferret auditory cortex to human spoken phonemes

Response of a ferret's primary auditory cortex to human-spoken words. The letters show the positions of neurons that best respond to that phoneme (characteristic sound of each letter). The response to the *plosive* sounds of *d, b, t, k, p,* and *g* are separate from the *fricative* sounds of *f,* and they each overlap somewhat with *z.*

Source: Nima Mesgarani. Used with permission.

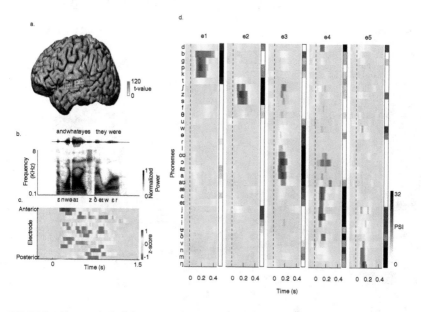

FIGURE 9.5 Human cortical response to human speech in the temporal cortex

Human cortical responses show selective responses to sounds in speech. (A) An MRI reconstruction of one participant's brain. The placement of electrodes during epilepsy surgery are shown in red. (B) An example spoken phrase, *and what eyes they were,* displaying sound waveform, spectrogram, and phonetic transcription. (C) Neural responses evoked by the different sounds within the spoken phrase at selected electrodes, with red showing the greatest response to that sound: anterior is toward the front of the brain and posterior to the rear. (D) Average responses for five example electrodes, labeled e1 to e5, to all of the phonemes used in English.

Source: Nima Mesgarani. Used with permission.

FIGURE 11.4 The Thai Elephant Orchestra

    Members of the Thai Elephant Orchestra performing on renaats and, in the standing percussion section, on a Thai temple gong, a gong made from an illegal confiscated logging saw blade, and a set of tuned tubular bells.

Source: Photos courtesy Millie Young, Mahidol University International College. Used with permission.

From our perspective, we can appreciate both Volta's invention of our contemporary world and also that Galvani was correct in claiming that animals produce electricity, thus founding the field of electrophysiology. Galvani even suggested that the oligodendrocyte cells in our nervous system insulate electrical fields, which accords well with our current understanding, *and* further predicted the existence of ion channels that convey electrical charge by neurons, insights that anticipated and underpin all of our understanding of nervous system function.

The Brain Produces Electricity

The definitive report that the human brain produces electricity was delivered by Hans Berger, of Jena, Germany, in 1924. Berger used as his experimental subjects his fifteen-year-old son, Klaus, and fourteen-year-old daughter, Ilse, inserting two electrodes under their scalps. He increased the power of the voltage difference—that is, he *amplified* it—and used the voltage changes to move a pencil across a paper chart recorder. Berger's recordings of brain's electrical signals from the head are known as *electroencephalograms* (EEGs), and the basics of the approach have changed little over the next century.

The signals from the children's heads were affected by what they were thinking or sensed. In 1932, Hans reported that when he asked Ilse to mentally divide the number 196 by 7 (she gave the right answer), as she began to concentrate, she displayed large electrical peaks at 10 Hz, a pattern still known as *alpha waves*. Similar alpha wave patterns occur around the visual cortex at the back of the head when you close your eyes. We also still use Berger's term for 18–25 Hz waves measured when one is awake, *beta waves*. Other researchers later described additional wave frequencies in the cortex's electrical activity, including delta (big slow waves when you sleep), theta, and gamma waves, each of which occur at a different rhythmic tempo.

EEGs are now typically recorded in the clinic by sixteen or more electrodes placed on the outside of the scalp. These record the activity of the cortex, the region of the brain closest to the skull. That EEGs effectively measure brain activity is astonishing when one considers that there is skin, bone, and additional layers of tissue between the cortex and the electrodes outside the head. Picturing this, you may realize why in order to see these waves, millions of neurons must be active simultaneously (figure 6.3).

The largest brain waves occur during sleep or anesthesia, when slow *delta* waves of about 2 Hz are prominent: it is as if the neurons are all holding hands, swaying together and singing "Kumbaya." At the opposite extreme, when one is awake and engaged with the environment, brain activity is composed of a range of smaller complex waves of many frequencies, including activity above 40 Hz, which are called *gamma* waves.

The same neuron can participate simultaneously in different rhythms, analogous to our discussion of how church bells and didgeridoos simultaneously vibrate at multiple fundamentals and harmonics. However, in contrast to the sound of a church bell, neuronal activity patterns are controlled by mechanisms beyond their intrinsic properties, particularly by inputs from excitatory, inhibitory, and modulatory synapses. Together, these factors produce an incalculable number of possible rhythmic patterns in the brain.

An important use for EEGs is to detect epileptic seizures and determine where they begin and how they spread to other brain regions. During seizures, a rhythm from one cortical region spreads too far to other regions. Rather than the normal awake state, which displays multiple simultaneous rhythmic patterns, during a seizure enormous areas of the brain fire together, similar to the delta rhythms during sleep, but faster. The complex patterns involved in thinking and awareness are disrupted. As we will discuss, there are rare people for whom these seizures can be triggered by specific musical patterns or styles, which apparently drive a pattern so strong that the rest of the nervous system is overcome.

Producing a Heartbeat with Batteries

For most neurons, activity occurs when they "fire" an *action potential*, which is a positive jump in their voltage, often by 100 millivolts (mV) or

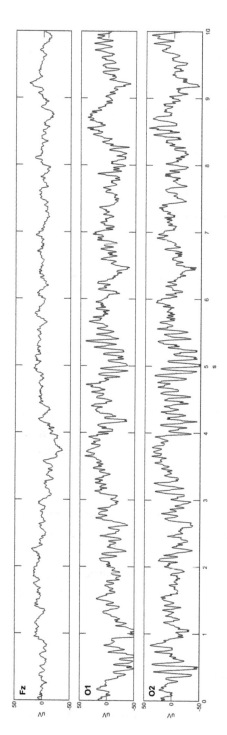

FIGURE 6.3 Alpha waves

Three electrodes simultaneously recording electrical activity of a subject's frontal cortex in the top trace and occipital (rear of the head) cortex in the lower two traces. The time in seconds is shown on the x-axis; changes in voltage are displayed on the y-axis. Prominent alpha waves are seen in the occipital traces around 4 and 5 seconds.

Note that different cortical regions produce different rhythms and that some peaks occur simultaneously while others are *phase shifted*, meaning that they occur before or follow a wave in another region.

Source: Recordings by Joseph Isler (Columbia University). Used with permission.

more, for about one-thousandth of a second (a millisecond: ms). In the brain, most neurons do not fire action potentials spontaneously and require activation from another neuron to initiate and sustain their activity. Some other cells, however, are spontaneously active, fortunately including those that produce our heartbeat: these are known as *tonically active* or *pacemaking* cells, as they maintain an internal clock.

The rhythm of the heartbeat is driven by the opening and closure of channels embedded in the pacemaker cell membrane. When the channels are open, electrical current flows between the cell and the liquid in which it bathes, the way that opening a faucet or channels in a dam allows water current to flow from higher to lower levels.

Analogous to the current of water molecules flowing downhill, electrical currents are carried by charged particles known as *ions* that move through these channels in the direction of higher to lower levels. When table salt is dissolved in water, it produces positively charged sodium ions and negatively charged chloride ions. To a biological cell, the difference in net positive and negative charges between the inside of the cell and the outside fluid are the two poles of a battery.

The pacemaking cells that produce our heartbeat live in the sinoatrial region of the right atrium. If you place a tiny voltmeter between the inside of the pacemaker cell and the surrounding fluid, you would measure a brief voltage jump from a *resting potential* of –70 mV to about +20 mV, at the top of the *action potential*, before the cell returns to the resting potential.

To understand the electrical currents of cells, you need only consider that like water, ions will move from a higher level to a lower level. The resting potential voltage is negative ("hyperpolarized") because of the opening of channels for potassium ions, which are positively charged. As there are more potassium ions in the cells than in the outside fluid, when a potassium channel opens, the potassium ions flow "downhill," in this case from the high concentration inside the cell outward to the low concentration in the extracellular fluid, thereby making the interior negative in charge.

When pacemaking cells are hyperpolarized, the low voltage activates the opening of *voltage-gated* HCN (H for *hyperpolarization-activated*) ion channels that allow sodium, the level of which is higher in the surrounding fluid, to enter and *depolarize* the cell. This depolarization then opens voltage-gated

T-type calcium channels, which allow calcium ions to enter and further depolarize the cell. As the concentration of calcium ions outside the cells is nearly 10,000 times higher than inside, there is an enormous potential to carry calcium ions into the cell, analogous to the powerful water pressure generated when our water tank is many stories above our bathtub.

Once the cell is depolarized to about −50 mV, it reaches a voltage *threshold* that activates the opening of L-type calcium channels. These are high-conductance channels, so the calcium rushes into the cell rapidly, producing an action potential peak of about +20 mV, the maximum voltage that pacemaking cells reach during their cycle.

As the voltage-gated calcium channels close, a set of potassium channels reopens to hyperpolarize the cell, "resetting" the cell to start the cycle again—and, one hopes, again and again, throughout one's life (figure 6.4).

As you know, the heartbeat is relatively slow, at rest beating about fifty to eighty-five times per minute, that is, a frequency of about 0.8 to 1.4 Hz. When pacemaking cells are not able to maintain a proper rhythmic cycle, a common surgical treatment is to implant an artificial "pacemaker" to apply the electrical current at the controlled intervals.

People with hypertension often take drugs (dihydropyridines) that partially inhibit the L-type calcium channels. This reduces the amount of blood moved by the heart and thus lowers blood pressure.

This opening and closing of ion channels to produce specific currents provides an ongoing rhythm of the heartbeat. It does not explain how the heartbeat is increased during exercise or slowed during rest: that is caused by modulatory transmission, a topic of the next chapter.

Why Neurons Are Batteries

Moving to the brain, why did Klaus and Ilse Berger's heads produce changes in voltage?

If you placed a tiny voltmeter to measure the difference between the fluid and the inside of a neuron, it would be typically rest around −70 millivolts.

Now let's excite that neuron and make it fire. This is typically accomplished by release of a chemical known as an *excitatory neurotransmitter*. The

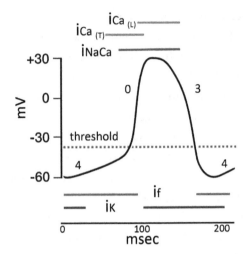

FIGURE 6.4 The rhythmic cycle of the pacemaking cells that control the heartbeat

The job of the pacemaking cell is to provide a rhythmic action potential that stimulates heart muscles to contract. Time is shown on the x-axis and voltage on the y-axis.

At *phase 4*, the pacemaking cell is at a hyperpolarized resting voltage, and there is a slow depolarization driven by a sodium current known as i_f (for "funny current"), caused by the opening of an HCN (H for "hyperpolarization-activated") channel. Since there are more sodium ions outside the cell than inside, this channel opening provides a slow influx of positively charged sodium ions.

Once the inward sodium current depolarizes the cell to the *threshold* voltage at the dashed line, T-type (for "transient") calcium ($iCa_{(T)}$) channels open. There are far more calcium ions outside the cell than inside, and so this channel provides an influx of positive calcium ions. There is also an activation of a protein known as a sodium/calcium exchanger (iNaCa) that trades 3 positive sodium ions (each with a charge of +1) for 1 calcium ion (which has a charge of +2), which further increases the depolarization.

At *phase 0*, there is a much stronger depolarizing current caused by the opening of higher-conductance L-type (for "long-lasting") calcium channels ($iCa_{(L)}$), which provides the entry for many more calcium ions, producing the peak of the action potential.

At *phase 3*, potassium channels (iK) open. There are more potassium ions inside the cell than outside, and opening these channels causes positively charged potassium ions to flow from inside to outside, returning the cell to a hyperpolarized resting potential until the cycle begins again.

(The nomenclature in the field doesn't necessarily make sense: there is no phase 1 and 2, and the channels and currents have multiple names.)

Source: Author.

neurotransmitter is released from one neuron, at the "presynaptic" side of the synapse, and binds to a receptor protein on the "postsynaptic" membrane of another neuron, triggering the opening of sodium-ion channels.

In contrast to the action potential peak of heart pacemakers, which are driven by the relatively slow entry of calcium, neurons open voltage-gated

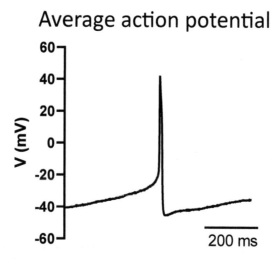

Average action potential

FIGURE 6.5 The rhythm of neuronal action potentials

An action potential recorded from an acetylcholine-releasing neuron (sometimes known as the giant neuron) in the striatum of a mouse. The activity of these neurons regulates other striatal neurons involved in learning. Striatal neurons are also affected by self-administered drugs such as nicotine, alcohol, and cocaine.

Source: Recorded by Sejoon Choi (Columbia University) in the author's laboratory. Used with permission.

sodium channels rapidly when the neuron is depolarized beyond a threshold of about –30 mV. The concentration of sodium in the extracellular fluid is about 150 millimoles (mM), while the concentration within the neuron is about 15 mM. The rapid influx of sodium produces an action potential that typically depolarizes the neuron by about 100 mV for about 1 ms (figure 6.5).

SIDEBAR 6.4

The Japanese puffer fish used to prepare *fugu* in restaurants has high concentrations of a poison, tetrodotoxin, that blocks the voltage-gated sodium channels on neurons' cell membranes. The chef needs skill when preparing this fish, as ingesting too much toxin blocks the ability to fire action potentials, causing asphyxiation and death.

Also like pacemaker heart cells, but faster, about the time that the voltage-sensitive sodium channels close, channels open for potassium ions, which flow from the higher concentration inside to the outside, hyperpolarizing the voltage and returning the neuron to resting potential. Neurons maintain a low internal sodium level by continually pumping sodium ions out from the inside of the cell to the surrounding fluid, against its *concentration gradient* (analogous to pumping water up from the bathtub to the water tank), typically coupling this energy-demanding sodium pump to the energetically favorable outward flow of potassium.

SIDEBAR 6.5

The terms "resting" and "action" potential are attributed to the German physiologist Julius Bernstein (1839–1917), who was following studies by Helmholtz on nerve conduction velocity and Walther Nernst on the flow of ions. Bernstein suggested the important role for potassium ions in maintaining the resting potential and speculated that neurons could alter their ion permeability, specifically by decreasing their permeability to potassium during the action potential. This conjecture was confirmed by a graduate student, David Goldman at Columbia University, during World War II.

The contemporary understanding of the role for inward sodium current in causing the peak of the action potential is credited to Alan Hodgkin, Andrew Huxley, and Bernard Katz, who at Cambridge University before and following World War II recorded currents from squid axons. Squids have a giant axon (about a millimeter in diameter, a thousand times larger than a typical axon), large enough to be impaled by the electrodes available at the time. To estimate the net current flow through the neuron at different voltages, one uses the "Hodgkin-Huxley" equations, which add together the contributions from each ion.

Neuronal Firing and Private Synapses

Neurotransmitter receptors that open sodium channels act to depolarize neurons. The most common of these *excitatory* neurotransmitters are glutamate and acetylcholine.

SIDEBAR 6.6

The food additive monosodium glutamate activates glutamate receptors that increase sodium ion entry, which causes overexcitation and headaches or fainting in some people. Some acetylcholine receptors that increase sodium entry are also activated by nicotine and so play a central role in the effects of tobacco.

Neurotransmitters that open potassium channels that allow positive ions to flow from the neuron outward, or that open chloride channels that allow negatively charged chloride ions at higher concentration outside to flow into the neuron, tend to decrease neuronal activity. In mammals, the most common *inhibitory* neurotransmitter is a glutamate metabolite known as GABA. Many commonly used drugs, including sedatives and antianxiety drugs, and possibly alcohol work by altering GABA receptor function. If your perception of time or response to stimuli seems slower with these drugs, their effects on your neuronal clocks may be the reason.

SIDEBAR 6.7

The discovery of neurotransmitters and receptors occurred during the era that radio was developed, with the first long-distance broadcast by Gugliemo Marconi in 1896. Radio uses a *transmitter* to broadcast a signal to home radios known as *receivers*, and so the physiologist John Newport Langley in 1905 used the terms *receiver* and *receptor* for the neuronal structure that was activated or "transduced."

Also during that period, the term *transmission* was used by Otto Loewi and Henry Dale, a student of Langley's, to describe the release of chemical signals from nerves. Lowei said that the idea for the experiment to prove the existence of neurotransmission using two dissected frog hearts, so that a signal from a nerve activating one heart could later diffuse to stimulate the other, came to him in a dream, and he rushed to perform it in the middle of the night in his laboratory.

During this same period, Marie Curie coined the word *radioactivity*, again in analogy to the radio.

(continued)

SIDEBAR 6.7 (CONTINUED)

The first neurotransmitter receptor to be thoroughly characterized, by Jean-Pierre Changeux and others, happened to be from the torpedo electric fish, whose electric organ releases extremely high amounts of the neurotransmitter acetylcholine (discovered by the aforementioned Otto Loewi). The very high levels of the acetylcholine receptor responsible for the electrical currents carried by sodium ions cause the fish's electrical shock.

The mechanism by which neurotransmitters are released is chiefly by exocytosis. Brilliant research by Bernard Katz and colleagues in Cambridge, mostly recording at the junction between nerve and muscle, demonstrated that this release occurs in small packets he called *quanta*. The nearly contemporaneous invention of electron microscopy demonstrated the presence of small internal structures known as *synaptic vesicles*, first identified with the electron microscope by George Palade and Sanford Palay. These *organelles* contain high concentrations of neurotransmitters and fuse with the neuronal membrane at synapses, thereby releasing the neurotransmitters contained within, which then bind to receptors on other neurons.

Fast-acting synaptic connections that rely on ion channels are often called *ionotropic*, whereas the slower modulatory synaptic connections that alter the tempo of the neuronal rhythms are often called *metabotropic*. As ionotropic synapses typically activate a short-acting local signal between a single presynaptic and postsynaptic neuron, I call them *private synapses*. Metabotropic synapses, in which a single presynaptic site releases a long-lasting signal that diffuses to multiple neurons, I call *social synapses*: we will discuss the latter in the following chapter.

Producing Neuronal Clocks in Rhythm

Most neurons in the brain, including the cortical neurons in Klaus's EEGs, are not pacemakers but require excitation from synaptic inputs by other neurons. Many of these neurons form large networks that are activated together.

Such a network can sometimes be seen on an EEG when a listener hears a regular beat. The neuronal waves are analogous to the water, sound, air, light, string, and other waves we have discussed, with the crest coinciding with the auditory beat at the frequency of the music.

The electrical rhythms of the brain in physiology laboratories are often detected by sound. The simplest approach is to adopt Hans Berger's technique and amplify the voltage that the electrodes measure but in this case send the signals not to an oscilloscope or a vibrating pen but to a speaker. My lab uses these sounds in physiology experiments to notify us if neurons are active. The sound isn't considered particularly "musical," as the rhythms we record are too slow (often 1–10 Hz) and irregular, that is, nonperiodic, to be perceived as musical notes. They sound like a series of clicks.

Listening to the sounds of firing neurons is clinically important, however. During surgery for deep brain stimulation, a treatment for Parkinson's disease that has now been conducted in hundreds of thousands of patients, a neurophysiologist listens to the rhythmic patterns to determine where in the brain the electrode should be implanted.

To study how the brain's electrical activity changes with rhythm, one approach is to play a simple rhythm multiple times until a steady pattern is clear to the listener and then to omit a beat, or delay it, or make it arrive too early. When the rhythmic irregularity occurs, there is a short burst of high-frequency gamma-wave activity of 40 Hz and higher that appears to reset the cortical firing patterns to a different pattern. This gamma-wave burst, called an event-related potential (ERP), is challenging to measure, and the experiment often needs to be repeated many times to confirm its presence.

In addition to changes in musical rhythm, a classic way to invoke an ERP is unexpected syntax—"the moon jumped over the cow"—I just triggered an ERP in your cortex. The presence of ERPs indicates that a series of expectations has been violated. It is likely that ERPs help us reset our neuronal clocks for musical patterns and tempi and inform us when we make an error in our motor activity, such as tripping while we walk.

When an ERP occurs during an EEG recording, we are recording waves of action potentials from massive numbers of both excitatory glutamate-releasing and inhibitory GABA-releasing synapses in the cortex. The role of the inhibitory neurons is probably to control the timing of the waves.

When an expectation in the music is violated, ERPs often occur between 100 to 250 msec after a "prediction error" (maybe a little later in infants, as in figure 6.6). This can be produced by an out-of-tune note, an unlikely harmony, or a surprising change in an otherwise predictable melodic pattern.

Perhaps a listener produces many ERPs while hearing of *The Rite of Spring* the first time but fewer on repeat listening?

ERPs seem important for maintaining and changing musical rhythms. Sylvie Nozaradan and her colleagues in Belgium and Montreal had adults tap specific rhythms, say, two eighth notes followed by two quarter notes and then two eighth notes, at different tempos. Where did the ERPs occur? Right

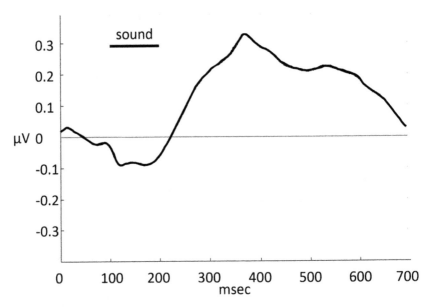

FIGURE 6.6 An auditory-evoked event related potential

An average of auditory event–related potentials (ERPs) evoked in infants by playing a 100 ms 750 Hz sound. Joseph Isler and colleagues recorded this response at the midfrontal cortex of thirty-four babies. The y-axis shows the change in voltage in microvolts and the x-axis the time after the sound is played, with a voltage peak that occurs about 300 msec after the sound is played.

Source: Recorded by Joseph Isler (Columbia University). Used with permission.

at the frequencies of the eighth and quarter notes. They concluded that beat and meter are both set by a series of ERPs.

They then sped up the tempo. At about 7 to 15 Hz, this simple rhythm became too fast to perceive (remember that at 20 Hz we begin to hear frequencies as notes), and the ERPs became weaker and messier. This suggests that ERPs are required for rhythmic perception and execution. It might explain why some tempos are too fast for much use in music—it might be interesting to test if drummers who perform the inhumanly fast snare rolls inspired by some genres of electronic dance music develop the ability to produce more accurate ERPs at rapid tempi.

The Nozaradan group noticed that in some cases, an additional series of ERPs would show up at exactly twice the frequency of the rhythm they were hearing, that is, at the first harmonic. This spontaneously occurring phenomenon may help explain the appearance of binary rhythms like clapping on the offbeat in gospel church music. Perhaps the establishment of new ERPs in the brain explains the ability to produce syncopation and to appreciate and play polyrhythms.

Interestingly, Istavan Winkler and collaborators in Budapest showed that ERPs occur in infants when their mothers bounce them on their knees but change the pattern by omitting a beat or adding syncopation. Perhaps this feature of rhythmic perception is widespread among species who listen.

Listening #6

To consider how Galvani's research affects the imagination, watch the 1931 film *Frankenstein*, starring Boris Karloff, in which the good baron reanimates his monster using Galvani's lightning rod and frog leg experiment, and then do follow it with the 1974 *Young Frankenstein*, by Mel Brooks and Gene Wilder.

There have been multiple attempts to use brain rhythms to produce music. The composer Alvin Lucier, rather than amplifying the waves and playing them through a speaker as in a neurophysiology lab, wired them to a device that struck a drum. He performed live concerts sitting on the stage and triggering the drums by wearing an EEG in *Music for Solo Performer*.

For the Brainwave Music Project, the electronic musician Brad Garton and I record EEG brainwaves during live performances and use software that he wrote to assign its features to make sound. For example, a higher frequency of EEGs might produce a higher note or louder volume. The signals can be separated by a Fourier analysis into component frequencies to produce harmonies or several lines of sounds simultaneously.

In that project, we sometimes have multiple performers play to see if their EEGs synchronize: since the musicians are hearing the same sounds, the sensory responses ought to be in sync, and if they are playing in rhythm, their motor responses should be, too. Nevertheless, a lesson from this project is that so much is happening in the brain that it can be a big challenge to observe simultaneous events simply from EEGs. The primary use for this EEG system has been for improvising musicians to perform live with their own brainwave-triggered sounds. A nice piece is "The Wheels," performed by the drummer William Hooker, and there is a live video of the pianist Rob Schwimmer creating a mini-concerto.

The unclassifiable composer, performer, and instrument inventor Raymond Scott created music intended to work as lullabies, as they were meant to resonate with the particular brain rhythmic activity of infants. Listen to his 1962 album *Soothing Sounds for Baby*.

Electronic drums performing seemingly impossibly fast drum patterns can be heard in Aphex Twin's (Richard James) "Flim"; Squarepusher's (Tom Jenkinson) "ken ishii x-mix," named after a DJ from Sapporo; and Venetian Snares's (Aaron Funk) "Kétsarkú Mozgalom." A drummer who has learned to play these rhythms live is Mike Glozier: listen to his "Venetian Snares."

Do you want your very own cerebral cortex to generate auditory-evoked potentials produced by listening to unexpected musical syntax? Just listen to Spike Jones and His City Slickers. Really, pick anything they perform.

7

• • • •

Neural Mechanisms of Emotion

• How does art affect emotion?

Determining the relationship between art and emotion is a daunting task, but interactions between fast-acting *private* synapses and slower modulatory *social* synapses are responsible for even more than the effects of art—they enable our interactions between our expectations and the environment and structure how we form our personalities, our speech, our beliefs, and how we will act in the future.

We'll start with hormones, then travel to the synapse, then to the enormous variety of drugs that work through these systems, and finally arrive at the crosstalk between private and social synapses that determine our interactions with the world.

Adrenaline, the Archetypal Hormone

Hormones are small signaling chemicals released from secretory organs: for example, insulin is released from the pancreas and regulates the level of blood sugar, and estrogen is secreted by the ovaries to regulate sexual characteristics.

Hormones circulate in the blood and transmit signals by activating receptors on other organs. This concept is very close to that of neurotransmission, and in fact some chemicals are both classical hormones and neurotransmitters.

Let's start with adrenaline, the first hormone identified, and like Hans Berger's invention of the EEG, it was said to have been discovered by a scientist experimenting on his child.

The adrenal glands are two relatively small organs encased in gobs of fat, one atop each kidney: hence, "ad-renal," meaning "next-to-the-renal" gland.

In 1895, the English physician George Oliver (1841–1915) discovered that an extract that he had prepared from sheep and cattle adrenal glands increased blood pressure: it was claimed by others that he made this discovery by administering the extract to his own son. Oliver noted that the rise in blood pressure was associated with a contraction of the arteries.

After Oliver's publication of his findings, the race was on to determine what component of the adrenal gland was responsible. Three different labs reported success in 1897. One called it *epinephrine*, which remains the most widely used term in the scientific literature. A competing drug company, Parke Davis, patented a preparation made from oxen or sheep adrenals under the name *adrenaline*. There has been so much commercial interest in adrenaline (for instance, it is now used as the "EpiPen" to treat severe asthma and allergic attacks) that it has been known by thirty-eight commercial names.

Consider what happens when you inject dog adrenaline into your child. The chemical moves through the bloodstream and binds to two major classes of receptors, *alpha* and *beta receptors*. In contrast to the receptors we discussed in the previous chapter, which rapidly open ion channels to produce electrical currents in neurons, hormone receptors activate cascades of enzymatic changes inside cells that can drive a multitude of responses.

Most of the modulatory receptors are now known as *G-protein-coupled receptors* (GPCRs). The term "G protein" refers to the activation of these proteins by a small signaling chemical known as GTP, a derivative of guanosine. Guanosine is one of the four nucleotides that compose DNA, but it also is an important intracellular signaling molecule in its own right.

The number of different GPCRs is stunning. Humans have a total of about 23,000 genes, and eight hundred of them code GPCRs! About 140 of these human GPCR genes are currently known as "orphan receptors," meaning that their activators are yet to be discovered.

The GPCRs are responsible for the effects of an enormous range of chemicals in biology, starting with insulin and adrenaline, as well as the well-known transmitters dopamine, serotonin, histamine, and melatonin; our nervous system's analogues of opiate drugs like morphine, known as the *endogenous opioids*; and the *cannabinoids*, which are our endogenous analogues of THC, the active ingredient in marijuana.

The tongue's receptors for sweet, sour, salty, and umami flavors are GPCRs. Even light is sensed by GPCRs, via the proteins opsin and rhodopsin, which in the retina transform photons into synaptic activity.

About 35 percent of used and abused drugs work by binding GCPRs. These include opium (and morphine, heroin, and Oxycontin), cannabis, caffeine, antipsychotics, LSD, and antihistamines. Other drugs such as cocaine, amphetamine, and certain antidepressants don't directly bind GPCRs but increase the levels of dopamine or serotonin and in that way act indirectly to increase the activity of GPCRs.

Given this extensive list of drugs, perhaps the claim that hormones and GPCRs are involved in emotion is starting to seem plausible.

The release of adrenaline into the bloodstream is mostly attributable to the activation of the awkwardly named *splanchnic nerve* by pain, anxiety, or trauma, which releases acetylcholine that stimulates receptors in the adrenal gland to release adrenaline and its derivative *noradrenaline* (a.k.a. *norepinephrine*) into the blood. Particularly by activating the alpha-adrenergic receptors found in arteries, this causes arterial constriction and increased blood pressure, as in George Oliver's son, enhancing the delivery of glucose, which supplies energy to muscles. This adrenaline-activated pathway is thought to underlie the stories of the amazing and unlikely feats of strength people perform emergencies, such as lifting up cars after accidents. It can also trigger tears (adrenaline and noradrenaline are found in teardrops), sweaty palms, trembling hands, and "butterflies in the stomach."

Do you recall that the regularity of heartbeat pacemaking is thanks to the currents conducted by the ion channels of the sinoatrial cells but that the changes in tempo are not? Adrenaline and noradrenaline are released from the adrenal gland, while noradrenaline is also released from sympathetic nerves that richly innervate the heart. The heart's pacemaking cells possess beta-adrenergic receptors, and these control calcium ions by activating their release from organelles inside the cells, contributing to the calcium-driven stages of the action potential. The initiation of the pace-making cycle is caused by these transmitter's effects on the "funny current," and these effects of adrenaline and noradrenaline on the rhythms of the heartbeat provide a paradigmatic example of why these transmitters are known as "modulatory."

Alpha-adrenergic blockers are used as drugs, including Trazodone, a widely prescribed antidepressant and sedative used to assist in falling asleep. In addition to slowing the heart's rate, *beta-adrenergic blockers* are used to treat migraine headaches and stage fright. You might consider from these effects that the archetypal hormone, adrenaline, can to some extent can affect the emotions.

Goosebumps

A challenge for the study of emotion is how to measure it. For example, how does one correlate brain activity with the appreciation of beauty? This challenge explains why there are studies on goosebumps, known as *piloerection*, a phenomenon we can identify as it occurs.

Goosebumps arise because tiny muscles known as *arrector pili* contract around hair shafts to form small depressions in the skin around the follicles, making the hair appear to "stand on end." While it would be nice to exercise your arrector pili muscles in the gym, the medical literature reports that as of 2018 only three individuals have been found who can produce goosebumps at will.

I'll give you a guess as to what triggers goosebumps . . . time's up . . . yes, it's adrenaline! The hormone activates the arrector pili muscles. A psychoactive plant extract, yohimbine, is an alpha-adrenergic agonist and also causes goosebumps.

Given that adrenaline has so many discrete effects, why do particular physical responses occur under different circumstances? To answer this question, we need to consider interactions within the nervous system.

Private and Social Synapses

Many of the paradigmatic physiological studies of synapses measured the synapse between the neurons from the spinal cord, known as *motor neurons*, and muscle cells. This synapse is called the *neuromuscular junction*. The signals flowing across these synapses cause muscle fibers to contract and are ultimately responsible for all of our voluntary movements.

The neuromuscular synapse must behave reproducibly and rapidly enough to provide fine motor control. The synapse relies on the release of acetylcholine from the motor neuron that activates the fast-acting nicotinic acetylcholine receptor on the muscle fiber (like the one characterized in torpedo electric fish), which opens sodium channels and drives an action potential in the muscle that lasts for 1 to 4 milliseconds.

The operation of the pitch-receptive neurons in the ear ("hair" cells), as we will see, likewise send fast-acting synaptic information through the auditory nerve, and a few rapid synaptic connections later, the signals triggered by hearing arrive at rapidly acting synapses in the auditory cortex. These features are required to perceive the fine and fleeting aspects of sound, and they are all transmitted though fast-acting private synapses.

GPCR transmission, in contrast, exerts effects that are longer lasting, usually over the range of seconds, although in some cases their effects can change circuits permanently.

For example, the neurotransmitter *adenosine* binds to a GPCR to help you fall asleep, which obviously changes virtually everything you do. Caffeine is a blocker (*antagonist*) of adenosine receptors, thereby helping keep you awake.

Oxytocin is a neurotransmitter in the brain and a hormone in the periphery that activates a GCPR that triggers states of closeness and social bonding and provides a signal for mothers to nurse infants.

For the control of emotion and mood, the most studied modulatory neurotransmitters are *serotonin*, as the most used antidepressant drugs, selective serotonin reuptake inhibitors (SSRIs), work indirectly on serotonin (as well as dopamine and noradrenaline) receptors, and *dopamine*.

Serotonin binds to at least fourteen different GPCRs in humans. In addition to its modulation by SSRI antidepressants, these receptors are indirect targets for the "diet" drug fenfluramine and direct targets for psychedelic drugs that can cause hallucinations, including methylenedioxymethamphetamine (ecstasy), psilocybin, and LSD.

Dopamine is a precursor to adrenaline and noradrenaline and a phylogenetically ancient compound found in many plants and animals: for example, bananas exposed to air become black from the oxidation of dopamine. Dopamine in the brain was initially thought simply to be a precursor for noradrenaline and adrenaline, although the Swedish neuroscientist Arvid Carlsson (1923–2018) showed that it had effects on its own, and an antipsychotic drug used to treat schizophrenic patients, chlorpromazine, worked as an antagonist of dopamine receptors in the brain.

A popular impression, right, wrong, or in between, is that dopamine is responsible for happiness, love, and the appreciation of beauty. This notion originates in one of the most amazing experiments in neuroscience. In 1954, James Olds (1922–1976) at the Montreal Neurological Institute was attempting to study arousal behavior in a rat by stimulating the reticular formation in the brainstem, a region deep in the brain, with an electrode. The electrode, however, was not sufficiently deep when he tried applying the electrical stimulus.

> I applied a brief train of 60-cycle sine-wave electrical current whenever the animal entered one corner of the enclosure. The animal did not stay away from the corner, but rather came back quickly after a brief sortie which followed the first stimulation and came back even more quickly after a briefer sortie which followed the second stimulation. By the third time of the electrical stimulus had been applied the animal seemed indubitably to be "coming back for more."
>
> (OLDS, "COMMENTARY")

Olds rigged a lever that the rat could push to "self-stimulate" by triggering electrical current from the tip of an electrode implanted in its own brain (figure 7.1). If the stimulation electrode were inserted into the region he identified, rats would press the lever two thousand times an hour for more than twenty-four hours straight. Indeed, their rate of lever pressing could progress from a starting exploratory rate of about ten per hour up to

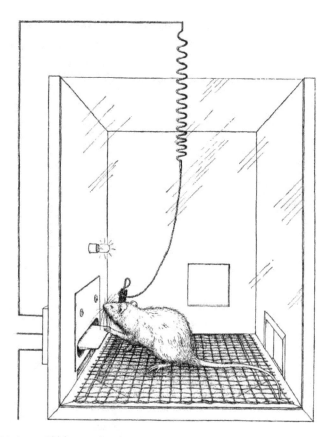

FIGURE 7.1 James Olds's reward pathway

After noting that the animal would return to the same area of the cage, apparently to receive the electrical stimulus again, Olds made a simple device so that the animal could administer its own stimulus by pressing a lever. He determined regions of the brain where an electrical stimulus was most "rewarding" or "reinforcing," that is, areas that would establish self-stimulation, and other regions where animals would quickly learn to avoid stimulating, that is, where the stimulus was "aversive." The region identified that caused the greatest increase in stimulation was a bundle of fibers that disinhibited the activity of dopamine neurons.

Source: Art by John Langley Howard, from James Olds, "Pleasure Centers in the Brain," *Scientific American* 195 (1956): 105–17.

five thousand presses per hour. Remarkably, he noted that this learned behavior was decreased by chlorpromazine, the antipsychotic drug that Carlsson was contemporaneously reporting blocked dopamine receptors.

Olds later altered the electrical stimulation lever so that animals could inject cocaine directly into local regions of the brain, which they would self-administer about ten times per hour.

For about a decade the neurotransmitter responsible for these behaviors was unclear. The initial research by Olds suggested that noradrenaline or a related transmitter was involved, and subsequent research from Roy Wise at the National Institute of Health, Gaetano DiChiara at the University of Cagliari in Sardinia, and others eventually settled on dopamine, the precursor to noradrenaline, as the major target. Indeed, all of the self-administered drugs that lead to addiction or dependence, including alcohol, opiates, nicotine, and amphetamines, have been found to enhance dopamine neurotransmission.

Dopamine Transmission Is Implicated in the Effects of Listening to Music

The evidence that dopamine neurotransmission underlies learned reward as in the self-stimulation experiments led to the widespread idea that "dopamine causes pleasure," but this remains a difficult concept to test in the laboratory, as we haven't really learned to measure pleasure. We can certainly measure increased dopamine release while animals have sex, drink alcohol, or eat chocolate and peanut butter, but it is now widely suspected that "pleasure" is caused by the release of endogenous opiates rather than by dopamine directly.

What we can observe is that increased dopamine release may enhance the number of repetitions of a task, a process known as *reinforcement*. Recordings of dopamine release in the rodent brain during learning by Mark Wightman and Regina Carelli at the University of North Carolina have demonstrated this relationship at the level of rapid changes that occur over seconds and even milliseconds. Experiments in monkeys during the performance of reinforcement tasks by Wolfram Schultz's lab in Cambridge show that the pattern of action potential firing by dopamine neurons follows the sort of rules that psychologists conjectured are involved in reinforcement learning.

The tools for measuring dopamine release in people, experimental subjects whom we can at least ask about emotions, are comparatively blunt. At present the field mostly relies on a brain imaging technique known as positron emission tomography (PET), in which a compound that binds to dopamine receptors is made radioactive. The compound competes for binding with native dopamine release; thus, the more dopamine released, the less space available for the binding of this PET ligand. This approach was used by Robert Zatorre and colleagues at the University of Montreal, who found that music that gives listeners "chills" is associated with greater dopamine release and increased blood flow in regions associated with dopamine release and that there is less dopamine release and activity in these regions when people listen to "neutral" music or "unpleasant" sounds. One might guess that something similar may be happening with endogenous opiate transmission, although this hasn't been ascertained.

Social and Private Synapses to Win Friends and Influence People

Even if the popular idea that more dopamine is related to pleasure seems fair, this can't be the whole story: why is it released, and how does it change brain circuits to reinforce and motivate future behavior? As mentioned, an action potential at a private synapse typically lasts a millisecond, whereas signaling by GPCR receptors responding to dopamine, noradrenaline, and other modulators at social synapses lasts for seconds and sometimes far longer. How can systems at such different time scales communicate? How can these interactions allow us to choose how to behave in response to the environment? These questions are being addressed in my laboratory and those of our colleagues.

The striatum is a region of the brain located under the cortex that integrates these signals, particularly from private synapses formed by cortical axons with dopaminergic social synapses. The role of these circuits is to integrate what we sense in our environment with our existing set of responses and choose from a set of competing possible behaviors. Some "loops" are chosen and others suppressed, and after a few more steps, the processed signals are sent back out to the cortex. The winning decisions are sent to the spinal cord to activate the selected sequence of muscles one by one during voluntary action. Even a simple behavior, like reaching for and grabbing an object, requires many muscles to contract at the right sequence, and this system in the striatum allows a proper sequence to occur.

These synapses can also change themselves, allowing us to learn new skills. As befits the breadth of activities we execute, there are billions of these cortical-dopamine-striatal synapses. While the number of midbrain dopamine neurons in humans is relatively low, about 300,000, they may each have up to a million synaptic release sites, in which case we may be considering a bewildering 300 billion possible synapses.

But as in late-night television ads, "but wait, there's more." When a release event occurs at a dopamine synapse, as a social synapse, it overflows to interact with many other synapses. If we examine a single private synapse at immense magnification with an electron microscope, we observe two inputs: a presynaptic input from the cortex at a private synapse with a "spine" of the striatal neuron and a nearby input from the dopamine axon forming a social synapse. The striatal neuron releases the inhibitory neurotransmitter GABA, and this also interacts with the other elements. There are additional nearby inputs from other striatal neurons and from other parts of the brain. We can call these "synaptic microcircuits."

At a microcircuit, each element has a receptor for the other transmitters and also a receptor for its own neurotransmitter, which is called an "autoreceptor" and typically feeds back to decrease the neuron's own activity. Even in this simplified version of a striatal microcircuit in figure 7.2, there are $2^{10} = 1024$ receptor on/off combinations at any time. Add to these the numerous other receptors and transmitters, the different durations of effects, and the many different inputs and synapses in this system, and the notion that these synaptic interactions provide us with so many possible behaviors—how to speak and move, what to say, and what to not do—may begin to seem reasonable. I tell my graduate students not to worry too much about running out of research questions in their future career.

So how can the slow-acting social dopamine synapse change the rapidly acting private synapse? In the striatum, the prevailing theory is that the dopamine release occurs after an unexpected "reward" from the environment, for instance, a performer realizing that the audience appreciates how she sang a particular note. The cortical signal that drove her to sing it

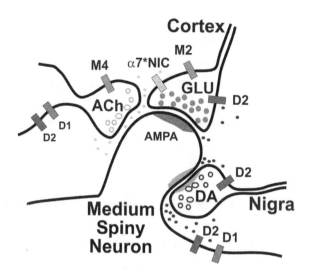

FIGURE 7.2 A striatal microcircuit

A illustration of a simplified striatal microcircuit indicates D1 and D2 dopamine (DA) receptors (these are on different popuations of medium spiny neurons); AMPA-type receptors for glutamate (GLU); and M2, M4, and alpha7 nicotinic receptors for acetylcholine (ACh). Even with only ten receptors indicated, this tiny circuit can exhibit $2^{10} = 1024$ states.

Source: Art by Nigel Bamford (Yale University). Used with permission.

in that fashion was recently active and has just sent a signal both to the spinal cord that activated the muscles that formed that note and a second *collateral* signal from an axonal branch to the striatum. In work by Nigel Bamford's lab at Yale, it appears that dopamine depresses the activity of those cortical axons that weren't involved in producing that unexpectedly successful note while leaving the synapses that were responsible for the success alone. The mechanism of inhibition appears to require the release of a "cannabinoid," a neurotransmitter that activates the same receptors as THC from marijuana, from a striatal neuron to feed back and inhibit cannabinoid GPCR receptors on the presynaptic side of the cortical axon. The cannabinoid release appears to occur when dopamine and glutamate activity are close to each other in time and space.

You can imagine how these changes at striatal microcircuits, thanks to the requirement of the coincidence of a successful motor command and an unexpected reward, might teach us how to adapt to our environment, learn to speak a language, or change how we will sing the next time.

Listening #7

There are multiple studies on the effects of music on blood pressure, with the selections reflecting the tastes of the scientists conducting the experiments—and for the experiments to be effective, the tastes of the subjects! There are published claims that J. S. Bach's Orchestral Suite no. 3 in D, which includes a movement rearranged as the well-known "Air on a G String," might be particularly effective for lowering blood pressure.

You will need to conduct your own research to find adrenaline-releasing art that is effective for producing goosebumps. While Camille Saint-Saëns's *Danse Macabre* seems to do it for many, I get noradrenaline-induced goosebumps and tear up with the Carter Family, Stevie Wonder, and Miles Davis and Gil Evans's *Porgy and Bess*. We all must be different.

This discussion of GPCRs provides an excuse to mention favorite music about drugs that affect GPCR activity. I am skipping songs about alcohol because, embarrassingly enough, the field of neuroscience still hasn't clearly identified its specific action or receptor in the brain. Here are some that we better understand:

"Spoonful," by Charlie Patton (dopamine receptors)
"Heroin," by the Velvet Underground (opioid receptors)
"Purple Haze," by Jimi Hendrix (serotonin receptors)
"Who Put the Benzedrine in Mrs. Murphy's Ovaltine," by Henry the Hipster Gibson (dopamine receptors)
"If Youse a Viper," by Stuff Smith (cannabinoid receptors)
Symphonie Fantastique by Hector Berlioz, widely suspected to have been written under the influence of opium.

I collaborated with the filmmaker Winsome Brown on *The Violinist*, a story of a Russian violin virtuoso of the Fritz Kreisler mode played by Rebecca Cherry. Her character becomes addicted to opium just as the drug was declared illegal by the Harrison Act in 1914. A piece intended to evoke the GPCR effects of opiates is "The Unfolding Opium Poppy."

8

• • • •

Ear Physiology

HOW AIR WAVES BECOME SOUND

- Does the peculiar shape of the ear change sound?
- How does the ear transduce air waves into electrical and neuronal activity?

Our Virgil for how the ear translates sound into electrical activity in order to transmit signals to the brain is Elizabeth (Lisa) Olson, a professor of otolaryngology (ear, nose, and throat) who studies hearing mechanisms in her lab at Columbia University. Some of her research examines the auditory system of the katydid, also known as the bush cricket, a singing grasshopper-like insect, and while their ears are underneath their knees, they operate in ways similar to ours.

Why Big Ears?

Our massive external ear funnels a large volume of air into the smaller ear canal, thereby increasing the pressure of the signal (figure 8.1). After the air waves enter the canal, there is still more amplification of air pressure as the canal narrows, so that sound wave amplitudes become much greater than in the outer air.

The folds of the outer ear, collectively known as the *pinna* or *auricle*, focus air waves toward the ear canal and in the process slightly delay and

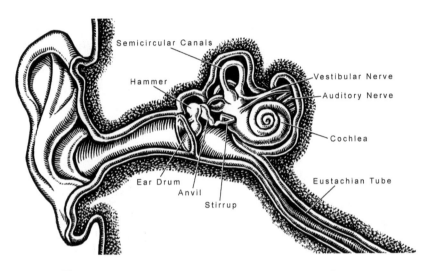

FIGURE 8.1 The ear
Source: Art by Lisa Haney. Used with permission.

defocus incoming sound. Like the body of a cello or guitar, the outer ear and canal resonate. The resonant frequencies are particularly powerful around 3 kHz, a bit above most fundamental frequencies in music but important for hearing the phonemes in speech.

The pinna further reflects and absorbs incoming sound waves, and this is thought to help decipher the location of the sound, including whether it comes from above or below the listener.

Our outer ear–to-ear canal ratio is not as impressive as that of house-cats, who not only have large outer ears but, like mini Linda Blairs in *The Exorcist*, can turn them 180 degrees to focus on a sound coming from a specific direction.

The Ear Drum

At the end of the ear canal, vibrations in the air encounter a small biological drumhead, the *eardrum* (also called the *tympanic membrane*). In an adult male, the area of the eardrum is about 60 mm^2 (a dime is about 250 mm^2). It is constructed of a thin layer of skin over mucosal and laminar layers.

The most influential scientist in this field, the Hungarian Georg von Békésy (1899–1972), examined dissected eardrums from human cadavers to demonstrate that the vibrations were indeed similar to the stretched membrane on a drumhead (figure 8.2).

At this point in the auditory pathway, the eardrum transduces airwaves into a different form of energy, mechanical vibrations, the way that air waves are transduced into mechanical vibrations by the membrane of a microphone. From here, the vibrations remain in the form of mechanical energy for several more steps.

The Bones of the Middle Ear

As shown in the drawing of the ear (figure 8.1), interior to the eardrum is the *middle ear*, a region filled with air that precedes the fluid-filled *inner ear*. At the border of the outer and middle ear, the eardrum passes its mechanical vibrations to the three smallest bones of the body, the *ossicles*. These are the *hammer* (malleus), *anvil* (incus), and *stirrup* (stapes, pronounced "staep-eez"): as you know if you make horseshoes, the hammer strikes the anvil. The ossicles vibrate with the mechanical version of the sound wave, and as the mechanical pressure at the eardrum increases and decreases, the ossicles move back and forth.

The final ossicle indeed resembles a stirrup, but rather than tapping the ribs of a horse, it raps on the entrance to the inner ear. The stirrup acts as a piston to transmit these vibrations to a membrane known as the *oval window* at the border with the inner ear, where the energy is transduced again, in this case from the vibrations of membrane and bones to vibrations of the fluid within the inner ear, known as *endolymph*.

Throughout this process of changing the mechanical means by which the energy of sound is conveyed, one might think that there would be a great deal of entropy and loss of the signal. Lisa Olson tells us, however, that while there is some reflection in the middle ear that dissipates sound energy at the frequencies where hearing is most efficient, about half of the sound energy that enters the ear canal is effectively transmitted from the middle ear to the inner ear.

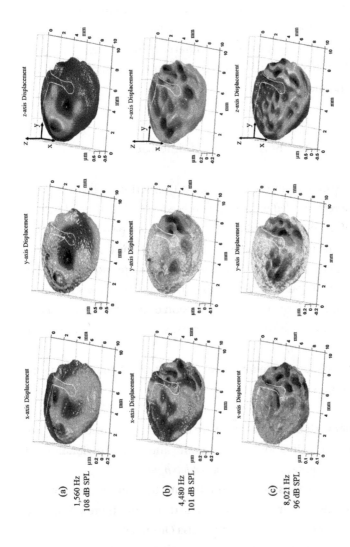

FIGURE 8.2 Vibration of the eardrum

In a contemporary version of von Békésy's approach, a human eardrum was stimulated at three different sound frequencies and the vibrations recorded by high-speed video. The rows represent three applied frequencies, and the three columns show the vibration in each spatial dimension, with red indicating the greatest and blue the least displacement of the eardrum membrane. The resonances resemble a snare drumhead.

Source: From Morteza Khalegi, Curlong Cosme, Mike Ravica, et al., "Three–Dimensional Vibrometry of the Human Eardrum with Stroboscopic Lensless Digital Holography," Journal of Biomedical Optics 20 (2015): 051028.

The *stapedius reflex* (also known as the *acoustic, middle-ear muscle, atten-uation,* or *auditory* reflex) occurs in response to loud sound and as one speaks. The stapedius, at 1 mm in length, is the smallest skeletal muscle in the body, and it acts to stabilize the stirrup. When the muscle contracts in response to loud sound, it pulls the stirrup back a bit from the oval window and decreases the transmission of vibration to the inner ear. When the stapedius muscle doesn't operate properly, for example in cases of Bell's palsy, this causes *hyperacusis*, in which normal sounds are perceived as being very loud.

The Cochlea, a Snail Filled with Seawater

The inner ear, which is shown in figure 8.3, consists of a coiled bony tube known as the *cochlea*, the Greek word for "snail," and, as mentioned, is filled with fluid. The length of the spiral stretched end to end, from the outer-most (*basal*) spiral to the innermost (*apical*) coil, is about 30 mm. The basal end eventually broadens and develops into large, striking, lobed Escher-esque *semicircular canals* of the *vestibular labyrinth*. These are responsible for the other major function of the ear, the sense of balance.

The tissue within the cochlea's spiraled tube is divided into three chambers. The uppermost is the *scala vestibuli* (or *vestibular duct*: *scala* and *scale* mean "ladder" and "stairs" in Italian), the middle is the *scala media* (or *cochlear duct*), and the lower is the *scala tympani*.

These three chambers are separated by two membranes. The separation between the *scala vestibuli* and *scala media* is the thin (two to three cell layers thick) *Reissner's membrane*. The *scala media* and *scala tympani* are separated from the lower chamber (*scala tympani*) by the elastic *basilar membrane*, which is our primary focus.

The scala media is filled with endolymph, which has a density similar to seawater. Unlike the other fluids of the body such as blood or cerebro-spinal fluid or, for that matter, genuine seawater, all of which are high in sodium, endolymph contains high levels of potassium. In this special case, where the normal ionic gradient is reversed, opening potassium channels carries potassium *into* instead of out from cells and causes a depolarization.

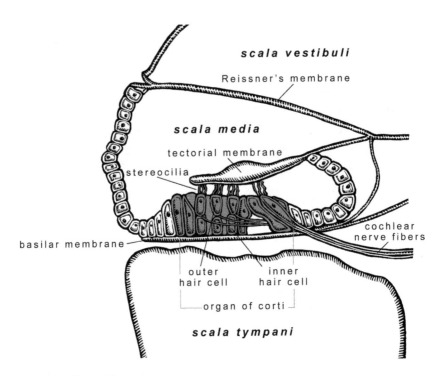

FIGURE 8.3 The cochlea

A section through the cochlea. The organ of Corti, shown in darker tone, lies between the tectorial and basilar membranes. The basilar membrane responds to higher frequencies at the cochlea's base and lower frequencies at its apex. Within the organ are the inner hair cells responsible for the detection of specific frequencies specified by their position on the basilar membrane. The outer hair cells are mostly responsible for local amplification of the mechanical responses to sound. The cochlear nerve fibers are responsible for sending the signals from the hair cells to the brain.

Source: Art by Lisa Haney. Used with permission.

The other two chambers of the cochlea are filled with the fluid *perilymph*, which is a more typical biological fluid, that is, one high in sodium.

Running above and along the length of the basilar membrane is a layer of epithelial cells known as the *organ of Corti* after the Italian anatomist the Marchese Alfonso Corti (1822–1876). Corti's organ is made from epithelial cells, including the *hair cells* responsible for sending sound information to the auditory nerve and from there to the brain. The organ also features an interior "tunnel" filled with perilymph. Yet another membrane, the gelatinous *tectorial* membrane, sits atop the organ of Corti within the endolymph of the scala media; it is important for the operation of outer hair cells.

The hair cells of the organ of Corti and its associated basilar membrane provide the steps in hearing that separate the sounds arriving from the outside world into their component frequencies so that they can be converted into chemical and electrical signals.

Lisa Olson explains the waves by which sound energy transfers from the tapping bones of the middle ear to the inner ear's structures:

> When the movement of the stirrup pushes the oval window in and out, this distends the basilar membrane. This distension moves the organ of Corti close to the stirrup in an up-and-down direction. That motion in turn launches a fluid and tissue wave of transverse motion down the length of the cochlea. At very low frequencies below our range of hearing, the fluid may be pushed longitudinally all of the way to the apex of the cochlea. At audible frequencies, however, the fluid moves only part of the distance.

So from these motions, a tissue inside of our ear that is only about 30 mm in length allows us to hear our entire frequency range. This should be impossible!

Recall that the wavelength of the note A4 above middle C (440 Hz) in air is 3.5 meters, about the height of two adults. Then consider that the speed of sound in seawater is nearly five times faster than in air at sea level (1500 m/sec versus 340 m/sec: see chapter 1). Multiply this out, and the note A4 in endolymph has a wavelength of about the height of a five-story building. The note A0, the bottom note of the piano, in the ear would have a wavelength of 5^4 = the height of a building 625 stories tall!

So how can an organ only 30 mm long provide all of the frequency information, across the entire range of our hearing, from about 20 to 20,000 Hz?

As you will now see, the term "organ" of Corti is appropriate not only as an anatomical term but in analogy to the keyboard instrument. Imagine the basilar membrane as a rubber band that becomes wider and thicker as it goes along: you know that thicker piano strings vibrate at a lower frequency than thin strings, so the skinny base of the membrane will resonate at higher frequencies than the thicker apex.

Now picture the band crossed diagonally with railroad tracks of tough fibers made from collagen, the protein that stiffens skin. The tracks are made with thicker and stiffer fibers at the base, yielding to floppier fibers at the apex that produce less tension. Recall that as with low piano strings, lower-tension strings vibrate at a lower frequency.

The outcome of this design is that the higher frequency components resonate at the stiff, thin basal end and the lower frequencies resonate at the larger, floppier apex. (The nomenclature is tricky for musicians, as the "base" of the organ of Corti is not producing the "bass" notes but is actually responsible for transmitting the more treble notes.)

You can imagine the isolation of resonant frequencies along the basilar membrane like the slats on a marimba, where larger slats vibrate at lower frequencies and each slat is isolated from the next, so that striking a B natural on a marimba transfers virtually no audible vibration to the neighboring A or C natural. In the organ of Corti, the stiffness of the basilar membrane makes the vibration fall off quickly, which means that the neighbors won't wake up. This design allows the component frequencies of a sound, like the components of sound observed in a Fourier transform, to be distributed along the "keyboard" of the organ (or marimba) of Corti like the notes of a chord.

Amazingly, the design of the basilar membrane was described in 1863 by the German polymath Hermann von Helmholtz (1821–1894), in his *The Sensation of Tone.*

> It is probably the breadth of the basilar membrane in the cochlea that determines tuning. At its commencement opposite the oval window, it is comparatively narrow, and continually increases in width as it approaches the apex of the cochlea . . . the radial fibers of the basilar membrane may be approximately regarded as forming a system of stretched strings, and the membranous connection as only serving to give a fulcrum to the pressure of fluid against these strings. In that case, the laws of their motion would be the same as if every individual string moved independently of all the others [like a piano]. . . . Under

these circumstances the parts of the membrane in unison with higher tones must be looked for near the round window, and those with the deeper, near the vertex of the cochlea.

OK, he's saying the same thing I did, but he wrote this a century and a half earlier—and more elegantly (I hate him!). What's more, Helmholtz determined from the number of fibers he could observe in the human cochlea that we should be able to distinguish about 4200 different musical pitches, which is extremely close to the current estimate. It would seem that Phill Niblock's five-hundred-pitch composition could be expanded upon.

So far, the auditory pathway has transduced air vibrations into mechanical vibrations in tissue and fluid and then separated the component frequency waves of the original sound across the specific resonant regions of the basal membrane. To send this signal into the brain, we need two other forms of transduction. The mechanical vibration of the basal membrane must be converted to electrical waves and then to chemical signaling at synapses. How these processes occur is the topic of most of the rest of this book.

Hair Cells Transform Mechanical Sound Waves to Electrical Sound Waves

The transformation of mechanical energy to electrical energy is known as *mechanotransduction* and is accomplished by the hair cells of the organ of Corti. "Hair" is an odd name for a cell type, but they appeared to possess "peculiar, stiff, elastic hairs," now known as *stereocilia*, to their discoverer, the German microscopic anatomist Max Schultze (1825–1874). Hair cells are used not only for hearing but in semicircular canals for the sense of balance. The human cochlea contains about 12,000 *outer* and 3500 *inner* hair cells.

The stereocilia arranged in *hair bundles* atop each of these cells consist of 20 to 300 individual hairs, with the rows sloping so that each successive hair is taller than the last, suggesting a Mohawk haircut. Each hair in a row attaches itself to the next taller by a tether near the cell's top, providing a means to make all of the hairs in a row move together.

Hair cells have synapses at their basal end, but in other ways they are unusual for neurons. In contrast to all of the other neurons we have discussed, they do not fire action potentials, they lack axons and dendrites, and they are sausage shaped, with their characteristic hair tuft at the top. They are typically classified as "epithelial" cells, a class that includes skin cells.

The base of the hair cells is within the organ of Corti, from which the body of the cell extends upward. The longer stereocilia of the outer hair cells reach to and attach strongly to the tectorial membrane. The upper part of the hair cell is exposed to endolymph, while the base is surrounded by supporting cells and bathed in perilymph.

The job of the hair cell is to respond to the mechanical movements of the inner ear and transduce those movements to electrical energy by opening ion channels. For hair cells to transduce sound accurately, they need to be able to release neurotransmitter at their synapses without an action potential being necessary. Consider that typical neurons require an action potential depolarization of 100 mV or so to trigger the release of neurotransmitter. Small changes in excitation don't activate typical neurons: their firing can be described as "all or nothing" or, if you think digitally, of ones and zeros.

In contrast, for seeing and hearing, we require a *gradient* of activation. Small changes in light and sound will slightly activate the cell. Larger changes will activate the cell more highly. A rod cell in the retina can detect a single photon, the lowest possible amount of light! Overall, the human eye can respond to about a billion-fold difference in light amplitude.

For hearing, the sound wave amplitude range we perceive can vary by about ten-million-fold: recall our discussion of loudness in chapter 1. These systems need to be *graded*, meaning that different levels of response can occur over a broad range of inputs. Indeed, increased transmission at the synapse between the inner hair cell and the auditory nerve neuron can occur at only about a thousandth of the voltage change that occurs during an action potential.

When the hair cell tufts are pushed and pulled by the sound-evoked movements of the basilar membrane, a mechanical force is produced that opens a motion-sensitive stereocilia ion channel. As the endolymph is uniquely high in potassium, these channels enable a potassium current to enter the cell, depolarizing it (in contrast to typical potassium currents that flow outward to hyperpolarize neurons). A calcium current also contributes to this depolarization. This is where sound is transduced from waves of mechanical vibration into waves of electricity.

The depolarizing currents from the hairs at the tip of the hair cell now activate voltage-gated calcium channels at the base of the cell. Depending on the size of this calcium current, synaptic vesicles—precisely how many is in dispute; it starts at one, but the postsynaptic response can range over twentyfold—can fuse to release glutamate as a neurotransmitter.

The hair cell synapses, mostly of the inner hair cells, activate the first bona fide neuron in our auditory pathway, the cells of the auditory nerve. Thus, the inner hair cells are mostly responsible for transducing mechanical energy to synaptic signals that from here on allow auditory information, now coded as electrical activity, to be conveyed through the brain.

Hair cells are present throughout the animal kingdom, but outer hair cells are only found in mammals. These cells play a role as mechanical amplifiers. When they are in motion, they produce tension because the taller hairs are fastened tightly to the tectorial membrane, so that the upper and lower membranes can move in opposite directions. The push and pull of the hair tufts produce a mechanical force that opens ion channels that drive depolarization by entry of positively charged calcium and potassium from the endolymph.

This mechanical amplification by the outer hair cells can occur because their membranes contain millions of molecules of a protein known as *prestin*, discovered by the physiologist and sculptor Peter Dallos and colleagues at Northwestern University. Prestin is thought to act as a piezoelectric material, meaning that it transduces mechanical and electrical signals by changing shape. The prestin molecules are embedded in the membrane on the sides of the outer hair cells. When the hair cell is depolarized, the prestin molecules are physically compressed in the membrane, making the membrane area smaller, so that the cell shortens. This mechanical vibration of hair cell length can be seen under a microscope when one plays sounds, and there are videos of hair cells apparently dancing to music. This outer hair cell–driven extra vibration at the correct region on the basilar membrane moves the membrane further at the right spot, thereby enhancing the sharpness of the frequency response by the neighboring local inner hair cells. The importance of outer hair cell amplification is seen in that prestin mutations cause deafness.

SIDEBAR 8.1

Piezo electricity was discovered by the brothers Paul-Jacques Curie and Pierre Curie in 1880 while studying the electrical properties of crystals such as in table salt and sugar. Piezo microphones are now widely used to amplify musical instruments during live performance, as they don't respond to vibrations in the air and are less prone to feedback and transmitting unwanted sounds. For string instruments, piezo mics are made from ceramic materials. They are often mounted on the bridge and so transmit little of the sound from the body. While proteins like prestin with piezoelectric properties are unusual, other biological materials can be piezoelectric, including bone.

Cochlear Otoacoustic Emissions

While the ear's job is to hear sound, the cochlea also produces sounds, sometimes loud enough to be heard by others. These sounds are known as *otoacoustic emissions*. They are thought to occur spontaneously in at least half of the population, although most of us are unaware of them when they occur. They are not auditory illusions and can be recorded by microphones.

Otoacoustic emissions can be evoked by playing clicks or tones into the ear, which are thought to activate outer hair cells to produce sufficient energy to move the basilar membrane. One approach, known as a *distortion-product otoacoustic emission*, is to play to two frequencies, typically between 800 and 8000 Hz, with one about 20 percent higher than the other. If the inner ear is functioning normally, it responds by emitting the difference frequency. Since otoacoustic emissions require an intact inner ear structure, this response is used as a noninvasive test for hearing, particularly in newborn babies.

Listening #8

Jonathan Ashmore from the University College London's Ear Institute made a video of an actual outer hair cell dissected from the low-frequency apical end of the cochlea that "dances" to a recording of Bill Haley's "Rock Around the Clock." You can watch the low frequencies activate the prestin piezoelectric response in the cell membrane to drive the hair cell to alternate between longer/thinner and shorter/fatter, in a rock 'n' roll rhythm.

James Hudspeth at Rockefeller University produced a cartoon animation of how the entire cochlea is thought to respond to the fundamental frequencies of Bach's Toccata and Fugue in D minor.

Maryanne Amacher's piece "Chorale" was written with the intention to produce otoacoustic emissia as well as combination tones within the ear.

9

• • • •

Deep Brain Physiology of Sound

- How do we tell the direction that a sound comes from?
- We only have two ears: how can we follow a specific conversation in a noisy cocktail party or a single instrument in a band?
- Why are we generally unaware of the sounds of own breathing and chewing?
- How do we associate music with times, places, and dreams?

Let's retrace the steps of sound's voyage into the brain:

1. The mechanical vibrations of *air waves* are amplified by the aerodynamics of the outer ear.
2. The amplified air waves are transduced to *mechanical waves* of vibrating tissue, initially at the eardrum like a microphone membrane and then to vibrating ossicle bones.
3. The final ossicle, the stirrup, drives the oval membrane to stimulate the liquid-filled chambers in the cochlea, producing *fluid waves* that drive the basilar membrane to vibrate.
4. Because of the progressive increase in size and decrease in stiffness along the length of the basilar membrane in the cochlea, the constituent frequencies of the sound resonate at specific locations, from the highest pitches at the membrane's base to the lowest at the apex.

5. These local waves on the basilar membrane create shearing forces at stereocilia at the tip of local outer hair cells: this works like a piezo microphone to compress and lengthen the cells, amplifying the local membrane waves.

6. The amplified local waves on the basilar membrane activate mechanosensitive ion channels on inner hair cells, driving calcium into the cells.

7. The calcium influx triggers synaptic vesicle fusion, which releases glutamate from the inner hair cell onto receptors on the postsynaptic neurons of the auditory nerve.

From this point forward, the journey of sound perception remains within the brain. Once there, the synaptic/electrical sound waves are sent to many regions, allowing us to listen to, create, and imagine sound and music. Indeed, neuroanatomists often call the synaptic pathways that descend from the auditory cortex to the rest of brain *corticofugal* in homage to the complexity of the musical fugue. This area of research is very active, and the more we search for synaptic connections, the more we find.

The Auditory Nerve Broadcasts on 30,000 Channels

We last left the auditory pathway at the synapses between the inner hair cells and the neurons that form the auditory nerve. In humans, there are about 30,000 axons in each auditory nerve, and we have one nerve per ear.

These neurons have an unusual anatomy. The cell bodies are in the *spiral ganglion* within the cochlea and in contrast to the typical design of one axon and multiple dendrites, they are considered to have two axons, a short *peripheral* axon that receives the synaptic inputs from hair cells and a far longer *central* axon. The central axons of these neurons all bundle together to form the *auditory nerve*.

Amazingly, electrical recordings of the auditory nerve firing can be played through a speaker and recognizably reproduce the sounds that entered the

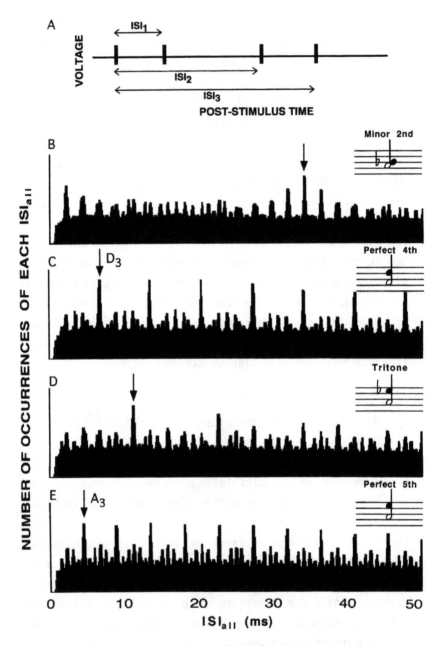

FIGURE 9.1 Auditory nerve response to musical intervals

Just-intonation intervals (Chapter 2) were played into the ear of a cat while recording action potentials from fifty axons in the auditory nerve. The consonant "perfect intervals" of the 4/3 perfect fourth (A4 and D5) and the 3/2 perfect fifth (A4 and E5) were compared to the more dissonant 16/15 minor second (A4 and B♭4) and the 45/32 (close to 7/5) tritone (A4 and E♭5).

ear. Such recordings indicate that the activity of auditory nerves can faithfully follow a sine wave up to about 4000 Hz, about the upper limit of the fundamentals used in music. Indeed, recordings of auditory nerve activity in cats suggest that other mammals may perceive consonance and dissonance similar to the way we do.

Recall that we perceive frequencies as high as 20,000 Hz but that even very fast neurons rarely fire faster than 200 Hz: while the neurons of the auditory system exhibit the fastest firing rates of any in the body, most tend to peter out at around 400 Hz.

The problem of how to transmit the firing of faster sound waves appears to be solved because a single hair cell can activate tens of auditory nerve neurons. This redundancy allows some of the 30,000 neurons of an auditory nerve to fire at some point during the wave peaks (figure 9.1).

Sound Reaches Its First Synapse in the Brain

The region of the brain that receives most of the synaptic output from the auditory nerve is known as the *cochlear nucleus*, an unfortunate name, as it is not in the cochlea but in the brainstem. The output axons from the auditory neuron map onto this nucleus in a high-to-low frequency gradient, so that the pattern of the basilar membrane's frequency distribution is

---

To display the response of the auditory nerve to the musical intervals, the durations between action potentials from each axon, known as *interspike intervals* (ISIs), are displayed. The x-axis indicates the ISIs in milliseconds, and the y-axis shows number of times each ISI occurred.

Remarkably, the auditory nerve ISI response featured large peaks of prominent *unplayed* frequencies that corresponded to f_1 fundamentals that would produce the higher note intervals. For example, when the cat heard a 3/2 perfect fifth, a prominent peak appeared that corresponded to the f_1 fundamental A3 220 Hz (indicated by the arrow at 1/220 Hz = 4.5 msec) that would produce an f_2 of A4 and f_3 of E5. When the tritone was played, a peak appeared at 88 Hz (11.4 msec: see arrow), which if it were an f_1 would produce an f_5 of A4 and f_7 of E♭5.

Note that "consonant" intervals exhibit larger and more regular components in neuronal activity than "dissonant" intervals, just as they do in sound waves (chapter 4), and that these perceptions apparently occur for other species.

Source: Mark Tramo, "Neurobiological Foundations for the Theory of Harmony in Western Tonal Music," *Annals of the New York Academy of Science* 930 (2001). Used with permission.

still preserved in the deep brain. Some of the auditory nerve's axons produce enormous synapses known as the *endbulb of Held* (after the German anatomist Hans Held, 1866–1942). These transmit signals to cochlear nucleus neurons known as *globular bushy cells*.

The axons of the globular bushy cells cross the midline of the brain to innervate a region on the other side associated with the opposite ear, the *trapezoid* body. This connection is the most rapidly responding synapse in the nervous system and is known as the *Calyx of Held*, after Held's impression that it resembled a flower's calyx. Like the endbulb of Held, it is extraordinarily large: in contrast to a typical synapse of about 1 µm (micron), the Calyx of Held is 20 µm across, the size of a typical entire cell body. This synapse is so large that the presynaptic side can be recorded by a thin wire electrode, and so the Calyx of Held has been valuable for the study of the fundamental properties of synaptic communication.

Marching to the Midbrain

Eventually, whether through a direct path or via an indirect path, for example through the Calyx of Held and the exotically named *superior olive nucleus*, the outputs of the auditory pathway mostly converge into a bundle of axons known as the *lateral lemniscus*, which extends from the brainstem into the midbrain. These axons synapse on neurons in the *inferior colliculus*, often referred to as the "midbrain sound center."

Nina Kraus's lab at Northwestern University found that if they place an electrode into the inferior colliculus of a guinea pig, the original sound can *still* be fairly well recreated by playing the electrical activity through an amplifier and speakers. This means that even at this advanced stage, the frequencies of the original sounds are transduced to similar frequencies of electrical activity in the deep brain. (It further suggests that at least up to this level, a guinea pig's sound perception is likely similar to ours.)

An important function of the inferior colliculus is to determine the direction of a sound in the environment. The way that small differences between the time that a sound reaches each ear are used to localize sound were characterized by Eric Knudsen at Stanford and Masakazu (Mark)

Konishi at Caltech, using the barn owl, a bird that relies on its ability to map the source of sound to find food.

As mentioned, a Calyx of Held synapse can transmit information at a very fast rate: some of these synapses appear to fire as fast as 1000 Hz, and some postsynaptic neurons fire as fast as 800 Hz without failing. Sound traveling from one side of a room takes nearly a millisecond to travel past your head, and so the more distant ear will hear the same sound a millisecond later. Because the Calyx of Held's synapses are so rapid, time differences between the arrival of sound to each ear (known as *interaural delay*) are compared when they arrive at the inferior colliculus, and the ability to detect these tiny delays serves to help us triangulate the direction from which the sound must have arrived.

Onward to the Thalamus

With the exception of smell, the pathways from each sense, including vision, touch, and the perception of hot and cold, converge into a large inner brain region known as the *thalamus*. For example, vision-related axons project to the *lateral geniculate* nucleus in the thalamus. For this reason, the thalamus is often called a "relay station" for sensory inputs on their way to the cortex.

For sound, the output axons from the inferior colliculus run major axonal output projections on both sides of the brain into a thalamic region known as the *medial geniculate nucleus* (MGN). Even at this advanced stage of the auditory pathway, the MGN maintains a frequency map, with the edges more responsive to low-frequency sounds and the center portions to high frequencies. The neuronal activity in the MGN is also controlled by the volume and duration of sound.

Daniel Polley's lab at Harvard reports that inputs to the MGN also arrive from the auditory cortex, a brain area that we haven't yet reached. This backward projection from a later stage in the auditory pathway produces what engineers call a *feedback* system. In this case, the cortical feedback can enhance or depress sensitivity to volume and frequency depending on the rhythm and timing of the cortical inputs.

This feedback loop appears to be important for deciding which sounds we notice or ignore. For example, in a noisy environment, the cortical feedback to the MGN may instruct us to ignore the synaptic pathways that would otherwise transmit the "background noise" and allow us to focus on the important aspects of the sound.

Victoria Bajo's lab at Oxford demonstrated that the feedback of the cortex to the MGN is important for determining if particular frequencies present in a sound are ones that would be absent from the normal harmonic series. This ability should help separate the overlapping sounds from specific sources, for example, finding which sounds are coming from someone speaking among many (the cocktail party problem) or distinguishing the instruments in a band.

It is striking is that we are mostly unaware of the sounds of our own chewing and breathing, although their sounds are loud in our ears. (Refer also to the stapedius reflex in the previous chapter, which occurs when we speak or sing.) Eric Bowman's lab at St. Andrew's University, in Scotland, suggests this is attributable to feedback from the cortex that inhibits the activity of another region in the thalamus, the *reticular nucleus*. It appears that this feedback, perhaps by inhibition of appropriate MGN neurons, confers the ability to ignore particular sounds and allows us to concentrate on those that are important. It is even suspected that the auditory hallucinations experienced by people with schizophrenia may result from an abnormal connectivity of these synapses, and this system might contribute to the sensory overload experienced by some individuals with autism.

The Auditory Thalamus Projects Into the Striatum

One of the MGN's major projections is to the *auditory striatum*. You will remember from chapter 7 that activities in the striatum are linked to emotion, learning, and choice. The auditory striatum receives dense input both from the MGN and the auditory cortex. How does the striatum integrate these signals?

A study from Qiaojie Xiong at Stony Brook University, in New York State, suggests that thalamic inputs from the MGN to the striatum act as an amplifier, whereas the cortical inputs to the striatum determine the frequency response. The coincidence of both inputs determines which will be effective and which will be ignored.

Another study from Mitsuko Watabe-Uchida at Harvard indicates that sounds can activate dopamine release in the auditory striatum, which could enable learning. For example, the sound of a train whistle might release striatal dopamine and encode a new synaptic circuit that associates that sound with a previously learned response to get off the tracks.

A clinical manifestation of the thalamostriatal auditory projections may occur in Parkinson's disease patients, who have trouble initiating movements but can still respond rapidly to a sudden sound like a car horn. This could be caused by thalamic activation of the striatal circuits that initiate a motor response even in the absence of normal cortical synaptic inputs. The circuit might help explain why Parkinson's patients with debilitating motor symptoms often are still able to dance well to music.

The Auditory Thalamus Projects to the Cortex

The MGN's auditory information also travels by massive axonal outputs to multiple cortical regions, particularly the *auditory cortex*. This region was discovered in the 1940s by the neurosurgeon Wilder Penfield (1891–1976), who left Columbia University's department of neurology in 1928 over academic power battles—some things never change—to cofound the Montreal Neurological Institute, where he conducted his most influential research.

Together with his colleague Herbert Jasper (1906–1999), Penfield introduced the *Montreal procedure*, a surgical approach to treat epilepsy. The procedure and its offspring are still used to treat epilepsy when drug-based approaches are unsuccessful.

After locating the general cortical region in which the seizure begins using EEG recordings, these operations identify the specific spot by applying a small amount of current to the electrode at different locations. The patient is awake during the surgery and can report their response. After the location is identified, because it triggers the characteristic seizure, the surgeon applies a bit more current to damage a small number of neurons in that part of the cortex. In most cases, the patient returns weeks later to the clinic with normal function and no further seizures.

The cerebral cortex in some mammals, including humans, is highly folded, increasing its surface area (figure 9.2). An inward valley is called a *sulcus* or *fissure*. The fissure that runs sideways across the temporal cortex on the sides of your head contains the auditory cortex and is called either the *lateral sulcus* or the *Sylvian fissure*, after the Dutch scientist Franciscus Sylvius (1614–1672). The hills that jut out from *sulci* (the plural) are called *gyri* (singular is *gyrus*), and the one that inhabits the auditory cortex is *Heschl's gyrus*, after the Austrian anatomist Richard Heschl (1824–1881).

Penfield's scientific writing is a pleasure to read, and he could have been an excellent mystery novelist—indeed, he wrote novels after retirement.

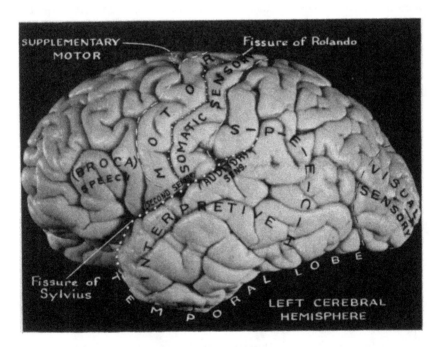

FIGURE 9.2 Wilder Penfield's diagram of a human left cortex

A photograph of the left cortex indicates the region of the cortex where electrical stimuli applied in the Montreal procedure produced sensory and motor responses. The region labeled "interpretive" produces "flashbacks" and experiences that resemble dreams, often including hearing sound and music. The motor, sensory, and visual regions and areas controlling speech are labeled, while the auditory cortex is located within the fissure of Sylvius and is not visible here.

Source: Photograph from Lister Oration at the Royal Academy of Surgeons of England, 1961.

His famous drawing of the "homunculus" illustrates the regions of the cortex that controlled muscle activity, which he naturally enough named the *motor cortex* (figure 9.3). When the appropriate point of the motor cortex is stimulated, the muscles in the corresponding body region contract, whether the patient wishes to move or not. Just across the central sulcus from the motor cortex are regions where stimulation produces sensation in specific regions of the body; he called that area the *somatic* (for *body*) *sensory cortex*.

While there are neurons that respond to sound in virtually all the lobes of the cortex, Penfield and Jasper, by stimulating regions of the cortex during the Montreal procedure, found the region that, from the verbal reports by the patients, was most responsive to the *perception* of sound. Penfield and

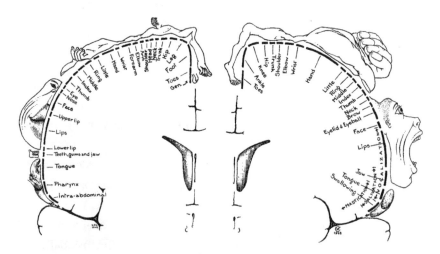

FIGURE 9.3 Wilder Penfield's cortical homunculus, a map of the sensory and motor cortex
    The central sulcus runs vertically to separate the sensory and motor cortexes: while they are next to each other, this drawing shows the parallels between the regions that control movement and receive sensory inputs, respectively. The large invaginations on either side below the faces is the lateral sulcus (a.k.a. fissure of Sylvius), which runs horizontally and separates the frontal and parietal lobes above from the temporal cortex below. The region around the lateral sulcus contains much of the auditory cortex (Brodmann areas 41 and 42, after the German neurologist Korbinian Brodmann, 1868–1918). The hill within the valley of the lateral sulcus, not labeled in this drawing, is *Heschel's gyrus* and contains the primary auditory cortex. The region where the lower temporal cortex juts out at the bottom of the drawing contains the "interpretive cortex" of the temporal lobe (Brodmann area 22) as well as *Wernicke's area*, required for speech comprehension.

Source: Art by Wilder Penfield (Montreal Neurological Institute).

Phanor Perot (1928–2011) reviewed these results in a 1963 review, "The Brains' Record of Auditory and Visual Experience." They relate that stimulation with their electrode deep in a difficult-to-access section of the cortex within the lateral sulcus, the *buried anterior transverse temporal gyrus*, "results in crude auditory sensation." Patients would say that they heard a tone, a buzz, or a knocking sound.

If the primary auditory cortex is damaged, even if the rest of the auditory pathway function is intact, subjects can lose the ability to be aware of sound.

This *primary auditory cortex* is fairly difficult to reach with electrodes, as it is deep within the lateral sulcus that separates the frontal and parietal lobes of the cortex from the temporal lobe below. Nevertheless, anatomists later divided the primary auditory cortex on the basis of synaptic connections and the presence of neurons that express specific transmitters into regions known as the *core, belt,* and *parabelt.*

Maps of the Aspects of Sound in the Auditory Cortex

The archetypal research on how the features of a sensory input are handled by the cortex is on the *primary visual cortex.* This region uses a set of rules revealed by Torsten Wiesel and David Hubel, who recorded from the cortex of cats while they watched flashing bars or spots. As you might suspect from teasing a cat with a laser pointer, the firing of neurons in their visual cortex is highly stimulated by light movement. Hubel and Wiesel found that the visual cortex is organized like a three-dimensional chess set, with some vectors (rows) of neurons firing in response to the different angles of the light bars and others to spots or bars that move in a specific direction.

The situation appears analogous in the auditory cortex, where electrode recordings in humans, macaque monkeys, and mice indicate that vectors of neurons respond to particular frequencies. Studies by K. V. Nourski at the University of Iowa suggest that specific neurons in Heschl's gyrus respond to specific sound frequencies and location. Likely, additional vectors are associated with "periodic" sound information, with neurons that fire, for example, when beats of pitches move closer to each other or with the repetition of a specific sound.

Sounds Can Be Reconstructed from the Activity of the Auditory Cortex

You'll remember that even after sound waves have traveled along the auditory pathway as far as the inferior colliculus, the sound can be reconstructed by playing the neuronal firing patterns through a speaker. By the time the auditory pathway reaches the primary auditory cortex, however, the features are too abstract, and playing back the firing through a speaker produces only clicks and buzzes. How does the abstract information in the vectors of the auditory cortex allow us to tell what we are listening to?

Some neurons in the core region of the auditory cortex appear to map to specific consonants and syllables. Daniel Abrams and Vinod Menon report that the core region of the auditory complex and striatum form a network by which infants can distinguish the sound of their mother's voice from the voices of other women.

Some more clues have been deduced from Nima Mesgarani's lab at Columbia University, who defined small regions in the ferret's auditory cortex that respond to specific human voiced phonemes, the distinct units of sound in spoken language (figure 9.4). This is amazing given that ferrets didn't evolve to comprehend human speech!

Moving to humans, in a contemporary version of the Montreal procedure for epilepsy surgery, an array of electrodes, rather than Wilder Penfield's single electrode, is placed on the surface of the temporal cortex. While the major goal is to identify where a seizure occurs, this also provides a means to determine which small regions of the cortex are activated by particular sounds. Mesgarani's lab played recorded conversations to patients while they underwent the procedure and recorded the response by the cortical neurons in the upper region of the temporal gyrus under the Sylvian fissure (figure 9.5). They analyzed the recordings from each electrode and

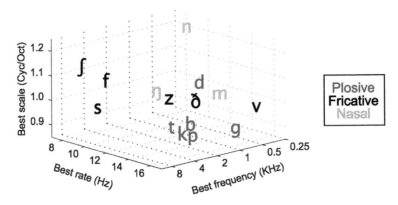

FIGURE 9.4 Neuronal response by the ferret auditory cortex to human spoken phonemes
   Response of a ferret's primary auditory cortex to human-spoken words. The letters show the positions of neurons that best respond to that phoneme (characteristic sound of each letter). The response to the *plosive* sounds of *d, b, t, k, p,* and *g* are separate from the *fricative* sounds of *f,* and they each overlap somewhat with *z.*

Source: Nima Mesgarani. Used with permission.

used an algorithm to analyze the response to each component sound (*phoneme*) in English speech. They eventually can use that information to reconstruct what the patient heard!

This research may eventually assist those with damage to the auditory system to regain language comprehension. Both Mesgarani's lab and Edward Chang's group at UCSF have been able to reconstruct speech from human brain waves recorded by electrodes touching the surface of or placed near the auditory cortex.

These new versions of the Montreal procedure are also beginning to unveil the perception of music. Mesgarani, Giovanni di Liberto, and collaborators have analyzed cortical recordings while playing fragments of J. S. Bach's music to patients undergoing surgery. They chose melodies with regular rhythms, including some melodic phrases that that were fairly predictable and others that were more surprising. They found increased cortical activity preceding notes that were just about to be played, demonstrating that the musical patterns were setting up expectations in timing. There must also be an expectation of pitch, as they found larger ERPs when the anticipated note was surprising compared to when it was predictable. (See Gordon Shaw's "Mozart effect," in the next chapter.)

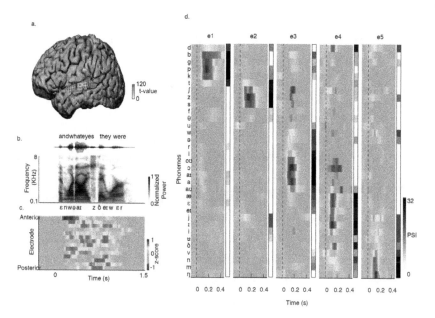

FIGURE 9.5 Human cortical response to human speech in the temporal cortex

Human cortical responses show selective responses to sounds in speech. (A) An MRI recon-struction of one participant's brain. The placement of electrodes during epilepsy surgery are shown in red. (B) An example spoken phrase, *and what eyes they were*, displaying sound wave-form, spectrogram, and phonetic transcription. (C) Neural responses evoked by the different sounds within the spoken phrase at selected electrodes, with red showing the greatest response to that sound: anterior is toward the front of the brain and posterior to the rear. (D) Average responses for five example electrodes, labeled e1 to e5, to all of the phonemes used in English. *Source*: Nima Mesgarani. Used with permission.

Each of the many parameters of sound, speech, and music requires a synaptic pathway for comprehension, and discovering and mapping these will be a rich and challenging field for a long time to come. There has so far been more attention given to speech than music, mostly in efforts to improve speech intelligibility after brain injury. However, Mesgarani's col-leagues have now been playing a broader range of musical excerpts to patients during surgery, and it seems that different musical sounds, such as violin sections versus hip-hop tracks, map to increased activity in par-ticular regions of the temporal cortex, reminiscent of the localization of phoneme responses in speech.

While there is a widespread assumption that our auditory pathways evolved for speech, one wonders, given the roles of music and other sounds in our and other species, perhaps this research will hint at how responses to sound, language, and music coevolved?

The Interpretive Cortex: The Stuff That
Dreams Are Made Of

Each cortical region projects to all of the others, most notably through the *corpus callosum*, a multilane highway of hundreds of millions of bundled axons. The cortical axons further project to the striatum, the cerebellum, and many other regions of the nervous system, including the spinal cord, for activation of voluntary movement. These myriad loops are involved in the pathways that underlie emotion, logic, intuition, deduction, and all of the other experiences in music and sound.

Well, if we weren't so complex, we probably wouldn't wonder how we think . . .

Another striking example of how the cortex is involved in sound and its interpretation again comes from Penfield and collaborators. They discovered a connection between the primary auditory cortex and a region just below it that they called the *interpretive cortex*. Penfield was so fascinated by the responses triggered in this region that he wrote an entire book on it.

The patients reported not only sounds, as if the primary auditory cortex were stimulated, but complete pieces of music or specific sounds and smells and associated scenes. Often these seemed to be reexperienced— the patients would often insist on this—but after the operation it became clear that they had never actually occurred; that is, they were not memories of a specific event in the patient's past. The perception of the scene would end abruptly once the surgeon stopped stimulating or moved the electrode.

The patients said that these experiences took place while they were aware that they were in the operating room, so that they experienced "two simultaneous situations" at once, and that both appeared real.

For example, for one patient (D.F.: at that time, the initials of the patient names were used in publications), when a point on the surface of the right temporal lobe was stimulated, she heard a particular pop song played by an orchestra. Repeated stimulations reproduced the same music. While the electrode was kept in place, she hummed the verse and chorus, accompanying the music she heard.

SIDEBAR 9.1

In the 1950s, the faculty at McGill and its Montreal Neurological Institute featured a large share of the personalities in this book, including Brenda Milner, Wilder Penfield, Herbert Jasper, James Olds, Peter Milner, and Donald Hebb. One of Penfield's collaborators on memory's localization in the brain during this time was Brenda Milner, known as the founder of the field of neuropsychology. As of this writing in 2019, she is at the age of one hundred still conducting outstanding research in her lab at McGill University in Montreal. Her husband, Peter, who worked with James Olds in the 1950s on the discovery of the reward pathway (see chapter 7), died in 2018 at the age of ninety-nine. An extensive interview of Brenda Milner by Heidi Roth and Barbara Sommer provides a wealth of information about that period, including her perspectives as a woman scientist during the mid–twentieth century.

Listening #9

Regarding the cocktail party problem in musical compositions, my vote for the award for the ability to include many lines of sound while retaining the ability to discern them as separate is Maurice Ravel's orchestration of *Une barque sur l'ocean*. Each instrument seems transparent, even within all the action evoking a boat rocking on the waves during a storm.

Henry Brant explored sound from multiple directions in orchestral works. You might try *Ice Field* played back as a binaural recording, a technique intended to recreate the sensation of sound arriving from multiple directions by carefully calibrating the timing to emulate the small differences between the times that sound arrives at the two ears.

Many musical traditions are used to evoke dreamlike states. The Gnawa tradition in Morocco, the Master Musicians of Jajouka in the Atlas Mountains, the ecstatic dances performed in a Sufi tradition for the whirling dervishes, and some Hasidic dances can be intended to evoke trances, as is some electronic club music (there is a genre called "trance"). Perhaps all music is intended to activate your interpretive cortex and produce something like a dream.

Some pieces, like Berlioz's *Symphonie Fantastique*; Karlheinz Stockhausen's *Gesang der Jünglinge*, with its abnormal manipulation of a child's voice; and Pauline Oliveros's *Bye Bye Butterfly*, with its disturbing setting of Giacomo Puccini arias, wear their hallucinatory intention on their sleeves.

The most awe-inspiring dreamlike states I have experienced with music and dance were in rumba ceremonies in Havana, where musicians, dancers, and audience appear to be entranced by an underlying rhythmic pattern for hours. There is no substitute for attending the ceremonies, but you might listen to Celia Cruz's recordings of prayers to the Orishas, to hear it from one aspect, and anything by Orlando "Puntilla" Ríos or Daniel Ponce, for another. I am told that Puntilla maintained a large Yoruba vocabulary hundreds of years after his ancestors had been kidnapped to be brought as slaves to the New World.

Since we typically filter out listening to our own breathing and swallowing, can we refocus on them as a form of music? Alan Watts recorded a lecture, "Listen and Breathe," that provides advice on being aware of the sound of one's breath as a form of mediation, and perhaps listening to it might alter some circuits in your nervous system.

10

. . . .

Sound Disorders, Illusions, and Hallucinations

- How does brain damage alter sound and music perception?
- Can epilepsy be triggered by music?
- What produces the auditory hallucinations associated with schizophrenia?
- What causes ringing in the ear?

In medicine and biology, insights into normal function are often made by studying disease or injuries. Here we examine examples of disorders, illusions, and hallucinations that lend insight into how we perceive sound and music.

Common and Benign Illusions

The Telephone Illusion

This is a universal illusion first described by the German physicist August Seebeck (1805–1849) using mechanical sirens. Telephones reproduce sound in a frequency range from 300 Hz (about a D above middle C) to 4000 Hz. In fact, recording engineers will apply a cutoff filter at those frequencies to provide a pretty good impersonation of speech through a phone.

The fundamental frequencies used in speech range in adult women from 165 to 255 Hz (middle C) and in adult males from 85 to 180 Hz. All of these frequencies are below the low-end cutoff and not transmitted. Yet we "hear" the absent fundamentals clearly over the telephone.

This illusion depends on the presence of the harmonics we discussed in chapter 3 and Mark Tramo's experiment in chapter 9. When we hear the higher harmonics of a fundamental, say starting at the $f_2, f_3 \ldots$, our cortex apparently fills in the fundamental f_1. For men with low voices, we only require the harmonics starting an octave or octave and a fifth above to regenerate the presence of the fundamental frequency.

Try recording your voice into a frequency analyzer—one free, downloadable program is Spear, written by Michael Klingbeil for his PhD thesis—and simply erase the fundamental frequency. It's surprising how much erasure of component frequencies one can do on the components of spoken or sung sounds before they become unrecognizable.

A long-standing question is how the cortex recognizes the fundamental pitch from a sound such as a cello or church bell, which produces many harmonics and overtones. One set of ideas, sometimes called "place theory," associated with Ernst Terhardt, speculates that there are templates stored in the nervous system that are compared with the incoming auditory information to determine the genuine or even replace a missing fundamental. Another suggestion, known as "temporal theory," is that the deduction comes from the brain's analysis of interspike intervals present in the auditory nerve's firing rates, as in Tramo's recordings in chapter 9. Of course, these are nonexclusive hypotheses.

The telephone illusion is not the only example of the brain filling in missing information. Normal vision includes a central blind spot, the *fovea*, in each eye, corresponding to a region in the retina that lacks photoreceptors. Objects in the blind spot disappear in our vision, but the region is filled in by a calculation by the brain based on the surrounding visual input. These auditory and visual illusions that fill in missing information assist us in navigating our world.

The McGurk Effect

This auditory illusion was described by the British psychologist Harry McGurk (1936–1998) and John MacDonald in the 1976 study "Hearing Lips and Seeing Voices." It occurs when there is a mismatch between an expected and genuine sound and is often demonstrated with a video of someone pronouncing a "p" sound but overdubbed with a "b." When one watches the video, the sound associated with the movement of the lips and face is perceived, but when an observer closes their eyes, the genuine overdubbed sound is perceived.

The McGurk effect depends on a learned expectation of a sound. There are disorders, including injuries to the cortex and learning disabilities, that inhibit the McGurk effect, and so in this case, the *absence* of an illusion may indicate a disorder.

The Tick-Tock Illusion

Also known as "subjective accenting," this illusion was reported by the American psychologist Thaddeus Bolton (1865–1948) in 1894, and you can perceive it by listening to a watch or to the turn signal in your car.

In his 1894 article "Rhythm," Bolton built a device to produce clicks through a telephone in the next room. When subjects listened to a series of identical clicks, they would initially hear them accurately as all alike. With further listening, however, they grouped the clicks into sets of two, three, or four, with one beat perceived as more prominent than the others.

In 2003, Renaud Brochard and colleagues from the University of Burgundy analyzed the tick-tock illusion using EEG recordings. They found that if clicks were identical, listeners would impose differences in ERP activity onto perceived strong and weak beats (for a reminder about ERPs and expectation, refer to chapter 6). When Renaud disrupted the perceived patterns by genuinely making particular clicks louder, listeners who were musicians tended to display larger ERPs, perhaps because they had stronger expectations of repeated rhythmic patterns.

Difference Tones

The violinist and composer Giuseppe Tartini (1692–1770) introduced the idea that two frequencies played simultaneously can produce a third tone that is the difference between the two. Called the *third sound* by Tartini and a *difference tone* since Hermann von Helmholtz, this is an illusion because this third tone is not present in the physical sound wave.

(This type of difference tone contrasts with difference tones produced by instruments such as distorted electric guitars playing two notes and by saxophone virtuosos who sing into their instruments while playing, as those tones are present in the sound signal.)

The perception of the difference tone illusion varies among individuals, and I tend not to perceive them unless the volume is quite loud and I really concentrate on distinguishing them. You might explore your own perception using a sound program to play a 1000 Hz and 1250 Hz wave and listen for a 250 Hz third frequency. For many, an effective approach is to hold one constant frequency and sweep the second frequency.

The biological origins of this hallucination are debated, but Fernan Jaramillio and James Hudspeth at Southwestern Medical Center in Dallas recorded from hair cells of the bullfrog and reported that difference frequencies arose within the hair cell itself. If these difference vibrations were then conveyed to the basilar membrane, they would excite hair cells at that resonant frequency. Other labs, however, reported difference tones even if one frequency is played into one ear and another frequency into the other ear, indicating that the difference tone is perceived at a later stage of the auditory pathway. Of course, both the ear and brain might be playing roles.

Can these kinds of auditory illusions be used to create new music from the inside of one's own head? The virtuoso of this compositional approach is Maryanne Amacher (1938–2009), who wrote music with high-pitched, closely packed waves intended to activate difference tones and distortion products, including the otoacoustic sounds produced in the cochlea we discussed in chapter 8.

Hearing Voices

It is surprisingly common to hear imaginary voices, particularly for children. While many who report these hallucinations show no other symptoms, it is by far the most common hallucination for people with schizophrenia and disorders associated with psychosis.

A possibly related phenomenon, including one I often experience, is to hear specific musical pieces in the mind. I have grown up thinking this was universal but am told that it is not!

A synaptic pathway regulated by the neurotransmitter dopamine is seemingly involved in hearing voices because drugs that block the D2 dopamine receptor, often called "antipsychotics," can decrease auditory and other hallucinations in patients with schizophrenia. While these drugs have been so used since the 1950s, the reason that they are effective is poorly understood.

A clue is that auditory hallucinations are reported to occur in tandem with a higher levels of activity in the auditory cortex and thalamus, pathways that both send sensory information to the cortex and striatum, suggesting that the hallucinations may be caused by abnormally low levels of synaptic inhibition.

One explanation is suggested by Stanislav (Stas) Zakharenko from St. Jude Children's Research Hospital in Memphis. Zakharenko studied a mouse with a genetic mutation also found in a fraction of patients with schizophrenia, and he observed that the mutant mice possessed higher than normal levels of D2 dopamine receptors on neurons that project from the auditory thalamus to the auditory cortex. This high receptor expression resulted in a disruption of information flowing from the thalamus to the cortex and also rendered these neurons sensitive to antipsychotic drugs. The administration of antipsychotic drugs restored normal auditory activity to mutant mice but had no effect on wild-type mice. These results hint at the possibility that people with auditory hallucinations might exhibit a dopamine-modulated disruption of communication between the thalamus and the auditory cortex.

Seizures Driven by Music

A related process may occur in rare patients with *musicogenic seizures*. In contrast to the widespread triggering of "goosebumps" by specific music, musical triggering of seizures occurs in perhaps one in a million people. EEG recordings show that in these subjects, music evokes a typical seizure that begins in one cortical region and spreads through much of the rest of the cortex.

These seizures are triggered by different music for different people, sometimes by a particular song and sometimes by a style or sound, such as "church music." One woman has seizures triggered only when hearing songs by the band Alabama. Musicogenic seizures can be effectively treated with antiepileptic drugs and in some cases by surgery, including the Montreal procedure.

There is a controversial hypothesis that Gordon Shaw's "Mozart effect," his theory that children would be smarter if they listened to and learned to play Mozart, can be used in the treatment of childhood epilepsy. As mentioned in the previous chapter, playing Bach recordings during Montreal procedure–like epilepsy surgery showed rhythms with increased activity that preceded predictions or expectations set up in the music. In some studies, the decrease in seizure activity lasted beyond the duration of the music, suggesting that such music may be therapeutic. While the evidence is unclear, we certainly have traced how sound and music travel to the cortex to modulate its synaptic activity, and in that way, playing a sound in a headphone or earbud is a noninvasive analogue to activating an electrode.

Tone Deafness and Amusia

A small fraction of people, generally from families with copious early exposure to music, develop "perfect pitch" (a.k.a. "absolute pitch"), which allows them to report the name of a musical note without a reference pitch. These people can be asked to sing, for example, an E♭ and nail it.

While early experience is clearly important, the reason that only some people have perfect pitch is not understood. One possibility is that the skill is more common than realized, as people will often sing a piece in the same key in which they learned it, with no reference pitch and without knowing the name of the key.

This possibility is consistent with a study by Robert Zatorre and colleagues from the Montreal Neurological Institute. They examined synaptic connectivity in people with perfect pitch. They concluded that these subjects have an improved ability to retrieve the name of the pitch thanks to enhanced synaptic connections between the right auditory cortex, which appears to be more specialized for hearing pitch, and areas in the left temporal cortex more devoted to speech.

A majority of people have "relative pitch," meaning that they can learn to recognize intervals and with practice and training determine a note from its relationship with a reference pitch. These people can, for example, sing a triad or scale correctly.

A small fraction of the population cannot learn these skills. The label *amusia* was introduced by the neurologist August Knoblauch (1863–1919) for those who are colloquially known as "tone deaf" and do not recognize musical pitch or are able to sing a melody in tune. Some tone deafness is "congenital" and present throughout life, but the loss of the ability to distinguish among frequencies also occurs following brain damage, particularly in the auditory cortex.

People with *congenital amusia*, a term introduced by Isabelle Peretz at the University of Montreal, maintain a normal understanding of speech and can recognize speakers by the sound of their voices, but they cannot tell if singing is out of tune or remember short melodies. They learn to identify songs from the lyrics. For some congenital cases, listening to music is unpleasant, but many people with congenital amusia are passionate music lovers.

The cause of lack of relative pitch is also not clear, but Peretz and colleagues using EEG recordings report that these individuals have reduced connections between the right auditory cortex and frontal cortex, the latter of which is involved in active "working" memory. This finding suggests that the steps in the auditory pathway work well until the stage of pitch perception in the auditory cortex, which could lead to less conscious awareness of the sound information. This condition may resemble a decrease in connectivity between the two sides of the brain that is suspected to occur in some people with dyslexia. Amusia and dyslexia may present somewhat analogous impediments in processing music or language.

Those with acute amusia usually have experienced brain damage to the auditory cortex, typically on the right side of the brain. This damage can occur through epilepsy, accidents, or from brain surgery. An extensive study was conducted on a patient known as GL who had an aneurysm on his right middle cerebral artery and after two surgeries was found to have two lesions, one in the left temporal lobe and one in the right frontal cortex. He previously enjoyed music and concerts, but after his surgery at age fifty-one, while his life continued normally and successfully otherwise, he noticed that he could no longer identify or enjoy previously familiar music. By testing for a range of discriminations of different musical parameters, Peretz found that he continued to discriminate among pitches normally but had lost the ability to follow the contours of melodies and could not sing back pitches that were played to him. Since he no longer enjoyed listening to music, these observations may provide a clue as to what is rewarding in music.

Tinnitus and Ringing in the Ears

The most common hearing disorder is tinnitus, defined as the perception of "phantom" sounds in the absence of a corresponding external acoustic stimulus. This is usually perceived as a buzz or hiss and often presents as a ringing sound at a specific frequency, most commonly around 4000 Hz.

While some estimates of prevalence report that 10 to 15 percent of people have tinnitus, with severe impairment in 1 to 2 percent of people, Lisa Olson finds that virtually everyone over age fifty has some level of tinnitus, defined as if they are in a quiet room, they hear sound that is not present. Noticeable tinnitus is particularly debilitating for musicians, as they must contend with a constant frequency while playing or listening.

As with deafness, there are many causes of tinnitus, and a specific diagnosis can be difficult. The causes of tinnitus further overlap with those of profound deafness, including noise, older age, tumors, and exposure to some drugs. It is associated with exposure to loud sound and so is unfortunately common among musicians and very common among those exposed to the noise of warfare. Often the region of the ringing matches the frequency where the hearing is lost.

The most common treatments for tinnitus are counseling and cognitive behavioral therapy. Some patients are helped by cochlear implants, hearing aids, and some newer experimental therapies, including brain stimulation. For many, the condition changes with stress and emotional factors, and some treatments appear to work by placebo effects. One musician friend with very debilitating ringing in the ear, which was destroying his career, improved significantly after he stopped smoking marijuana.

Some investigators state that the most common form of tinnitus is *somatosensory*, that is, associated with head and neck trauma that alters nerve inputs. This type often changes in volume, frequency, or localization of the ringing tone with altered head or neck position and can be altered by clenching teeth, turning the head, or applying pressure to the head or neck.

Somatosensory tinnitus may develop because of changes in synaptic inputs to the cochlear nucleus that affect its output. For some, the cause may be altered inputs from nerve fibers or jaw or neck injuries that change sound conduction. Experimental treatments include subcutaneous stimulation of these regions or controlled botox injections (the active toxin from the incredibly poisonous botulinum bacteria), which decreases muscle activity by severing proteins involved in synaptic vesicle exocytosis.

Some less common forms of tinnitus can be audible to the examiner and so are called *objective tinnitus*. This can be caused by altered blood flow in blood vessels near the ear or contractions of the eardrum. The most common type of objective tinnitus occurs in synchrony with the heartbeat and is caused by an impediment in transporting blood through veins and arteries. Other forms are related to spontaneous otoacoustic emissions.

Even if the initial cause stemmed from a problem in the ear, for many the tinnitus continues after the auditory nerve is sectioned, indicating a deeper brain pathology. Some reports hint that tinnitus may be associated with auditory remapping anywhere along the synaptic auditory pathway, including synapses from inner hair cells to the auditory nerve or from the cochlear nucleus to the cortex. Another cause is a decrease in inhibition of firing in the cortical tonotopic map of a region corresponding to the specific ringing frequency.

Mechanisms That Produce Deafness and the Effects of Loud Sound

People who do not hear well can find themselves cut off from music and challenged in social situations. We have discussed the many physiological steps required for hearing, and so it may not be surprising that over three hundred syndromes are associated with hearing impairment.

While there are many causes of deafness that cannot be avoided, a major one is simply exposure to loud noise, and noise-induced hearing loss is estimated to account for 40 percent of all occupational disease. Cumulative sound above 85 dB, which is quieter than some restaurants, seems to cause gradual damage, and obviously this is a major problem for musicians, particularly those who play loud instruments. Another preventable cause is chronic exposure to heavy metals or benzene.

The disorders of hearing loss are bundled into four major categories, defined by where the problems occur in the sound pathway.

Conductive hearing loss is a mechanical disturbance of the conductance of sound from the pinna through the ossicles. If the problem is in the external ear canal, sound is perceived as being quiet, whereas if it is at the ossicles or eardrum, there is generally a problem with frequency response, so that high or low frequencies are perceived as louder or softer than normal.

The causes of conductive hearing loss range from simple blockage by earwax, which is particularly common in children because the canals are smaller. Children also are more prone to infection, eardrum perforation, and a growth behind the eardrum known as a *cholesteatoma*.

As we will see in more detail later when we discuss hearing by whales, for whom sound is conducted through the head to their ears, some of our sound perception is conducted by bone, as you can test by striking a tuning fork and touching a bony area of your head. For this reason, damage to the jaw or regions of the skull also leads to conductive hearing loss.

Sensory hearing loss is caused by a dysfunction of hair cells or their synaptic connections to the cochlear nerve. This is commonly caused by a loss of outer hair cells and thus their mechanical amplification. This type of loss can be identified by a decreased response in the evoked otoacoustic emission tests that report outer hair cell function.

Sensory hearing loss is responsible for the common loss of perception of quiet sounds in older age (known as *presbycusis*) and for distorted perception. It is very commonly caused by noise, including the use of guns, explosions, and, sadly, very loud music heard in concerts or through headphones and earbuds.

Hearing loss from the sound levels at loud concerts appears to be attributable less to direct injury to the hair cells than to oxidative stress caused by a metabolic reaction to the sound, possibly following inflammation. This type of hearing loss appears to be greatest at 4000 Hz and higher. The best protection is obviously to avoid the concert or use earplugs, but if one is exposed, there are controversial claims that dietary antioxidants such as n-acetylcysteine may be helpful, perhaps by decreasing the levels of an inflammatory protein known as tumor necrosis factor. Several other disorders are now treated by tumor necrosis factor inhibitors, and it will be interesting to see if over time this leads to pharmacological measures for lessening the effects of age- and noise-dependent sensory hearing loss.

Hearing loss from age may further be attributable in part to loss of the microvascular blood supply for the hair cells or other cells associated with the middle ear. It can also be caused by infections of the ear, including mumps and measles, and by some antibiotics, including streptomycin and gentamicin. For some, sensory hearing loss has been attributed to the use of tobacco, cocaine, heroin, or alcohol. Finally, a quarter of cases of sensory hearing loss are thought to be caused by a large set of mostly recessive genetic mutations.

Neural hearing loss is chiefly caused by cochlear nerve dysfunction. This often causes a problem in perceiving speech rather than in frequency perception. It can result from an abnormal growth or a tumor, the most common of which is an *acoustic neuroma* or *vestibular schwannoma*. If you notice a loss in sound acuity in one ear, call a doctor immediately about undergoing an MRI scan. Fortunately, this form of deafness is often treated successfully with surgery.

Finally, *central hearing loss* occurs with damage to the cortex and has a host of causes, including infarcts, bleeding, tumors, and inflammatory problems that are consequences of other disorders, such as multiple sclerosis.

Treatments for severe hearing loss vary with the diagnosis, but the technology for cochlear implants has been advancing. For small children whose hearing is impaired by more than 90 dB, cochlear implant technology is responsible for vastly improving the ability to acquire speech. Implants have not yet proven successful for regaining the ability to perceive music.

The American composer Richard Einhorn was struck with sudden severe hearing loss in his right ear in 2010, perhaps from a viral infection that damaged the inner and outer hair cells. He reports that he can perceive music well again by the use of a hearing loop or "t-coil," a copper wire that radiates radio signals that can be picked up by hearing aids and cochlear implants.

Far more research on profound deafness and hearing loss is needed, but the improved understanding of sound perception in the cortex provides an optimistic outlook for future approaches and devices that will better enable prevention and treatment.

Listening #10

The quintessential telephone frequency cutoff in popular song must be Paul McCartney's "Uncle Albert."

A nice video demonstration of the McGurk effect—this has to be seen and heard—is made by the magician Mark Mitton, who has a deep interest in science. He produces only one sound, *ba*, and by dubbing it onto his mouth movements apparently speaks a different sound. If you both watch and listen, you might hear *tha*, *va*, and *fa*, but if you shut your eyes, you will perceive only *ba*.

Compositions that take advantage of the tick-tock illusion, which creates a perception of patterns from arbitrary/random timing, include Gyorgi

Ligeti's *Poème symphonique*, in which 100 metronomes set at different tempos are played simultaneously, and Douglas Henderson's *Music for 150 Carpenters*, with the performers using hammers, tool belts, and lunchboxes.

Some instrumentalists specialize in producing difference tones. On the tenor saxophone, one sings while playing to produce these. This technique is responsible for the growls of Earl Bostic (try "Harlem Nocturne," "Cherokee," and "Flamingo"), King Curtis ("Soul Twist"), and John Coltrane, Ben Webster, Lee Allen, Lenny Pickett, and Pharoah Sanders ("The Creator Has a Master Plan"). The sound of electric guitars distorted in tube amps can produce strong difference tones when playing on two strings. Jon Catler has a nice solo electric guitar demonstration of these phenomena using his microtonal guitar.

Provocative attempts to produce/receive inspiration from musical illusions can be heard in Maryanne Amacher's "Head Rhythm 1" and "Plaything 2." These pieces use different patterns on two stereo speakers and seem to produce music very different from what you would suspect from listening to the tracks separately.

The Sonata for Two Pianos in D major K448 by Mozart as performed by Murray Perahia and Radu Lupu is the typical work used to study the "Mozart effect." Gordon Shaw and his wife Lorna told me that this piece was first used because the boyfriend of Frances Rauscher, the first author of the original paper, had it in his record collection. Gordon, who introduced the "Mozart effect", was a personal friend who disliked contemporary composed music. I'm afraid he felt that minimalist music destroyed the ability to concentrate. I will not mention the pieces he played to examine his hypothesis, as it is pretty unfair to the genre!

There are multiple instances of great composers who could create wonderful work despite severe hearing loss. The most well known by far is Ludwig van Beethoven, who composed his last five piano sonatas, final five string quartets, the mass *Missa Solemnis*, and the Ninth Symphony while he is thought to have been nearly or completely deaf. Reports of the premiere of the Ninth state that he did not realize the audience was cheering until a singer turned him around to face them.

Bedřich Smetana was deaf when he wrote his best-known orchestral work, the symphonic poem *Vltava*, also known as *The Moldau*.

Gabriel Faure wrote his piano trio and only string quartet after being struck deaf.

Ralph Vaughn Williams, who lost his hearing from noise exposure in France while fighting in World War I, wrote his own Ninth Symphony in E minor after he became deaf.

The composer Richard Einhorn is also a recording engineer and has made demonstration recordings representing what sound is like with various levels and forms of hearing loss and with the use of specific tools such as hearing loops.

Evelyn Glennie is a composer and master percussionist with dozens of great recordings in a range of styles. She has been profoundly deaf since childhood. Her "Hearing Essay" discusses this issue in detail. One of the excellent pieces written for her is John McLeod's *Percussion Concerto*.

11

• • • •

Animal Sound, Song, and Music

- Is music limited to our species?
- Can other animals play musical instruments?

It seems important for our species to define ourselves as unique. This task shouldn't be too overwhelming: name another species whose set contains Groucho Marx (OK, Bugs Bunny). Our rules to defend our unique nature, such as "only humans use tools," or "only humans have a sense of ethics or tell lies," or "only humans maintain bank accounts," exist to be violated.

Most mammals, birds, and amphibians and some insects, reptiles, and fish produce short calls such as roars, chirps, and barks. These vocalizations occur on land, in the air, and underwater, in great choruses at dawn and dusk and during migrations. It is a challenge to perceive them as anything other than a fugue of many voices.

Bernie Krause, a musician and ecologist who records the sound of natural environments, introduced the idea that animals alter the frequency and timing of their vocal productions so that they are heard and not masked by others in the vicinity. He suggests that the health of a wild environment can be measured by examining its "soundscape" for the distribution of sound frequencies and that a healthy frequency distribution is damaged when human intervention has eliminated some of the sound band, resulting in broad gaps in the spectra.

Individual songbirds, baleen whales, bats, and gibbons produce songs so rich and complex that no one hazards labeling them as anything else. If we insist on being special because we make or appreciate music, we must explain away birdsong as "not music," and I haven't heard a cogent argument for that.

To appreciate animal music and sound, we need to overcome some barriers particular to our species. For example, recordings of brain activity in frogs and some birds indicate that their neural activity responds to songs by their own species far more than to other sounds, so that they appear to live in a hyperaware state of their cousins amid a cacophony of other less interesting sound.

As another example, because their vocal frequencies are above our range of hearing, we only recently realized that mice produce songbird-like songs.

It has only become recently known that there is a "dawn chorus" of fish calling in some regions of the sea.

We simply aren't present in the caves when bat species sing birdlike songs to one another or while flocks of millions of birds produce complex and interlocking, likely interacting, night-migration calls.

The biologist and naturalist Roger Payne calculated that before the era of noisy turbine-driven commercial shipping and mechanical drilling, some species of whales produced songs at frequencies and volumes that could be received by other whales across the span of an entire ocean. If this was so, whales were surrounded by an orchestra of calls and song throughout their lives.

Species besides our own sing to be attractive to mates; proclaim their individuality, location, and territory; to practice; for social bonding; to decide who belongs or is excluded from a community; and because they find solo or group singing rewarding. Are these reasons familiar?

How Birds Sing

The bird's vocal apparatus doubles ours in that the *syrinx*, named for the double pipe instrument played by the god Pan, produces two frequencies. This organ, found in all birds except vultures, is located further down the throat than our larynx, in the chest. There, the trachea branches into the two lungs, with two sets of muscles that control *labia* (from "lips") on either side. This anatomy allows birds to move very rapidly between pitches, and some, including the brown thrasher, catbird, and vireo, can sing two notes at once.

For an example of how the double pipes of the syrinx can be used, Roderick Suthers from Indiana University reported that the two-octave sweep in the song of the cardinal begins in the right branch of the syrinx at 9 kHz and then switches seamlessly at around 3.5 kHz to the left branch, to finish the phrase at around 2 kHz (figure 11.1).

Songbirds principally vocalize and hear mostly in the range of 500 Hz to 12 kHz, although hummingbirds produce sound at up to 16 kHz.

A celebrated singer is the European common starling. Mozart is said to have taught a theme from his Piano Concerto no. 19 to his pet starling. Stewart Hulse and collaborators found that starlings can distinguish rhythmic patterns by training them to respond by pecking specific computer keys for a food reward. (As we will discuss later, zebra finches can operate tiny musical instruments to trigger sounds they want to hear, including songs of other birds and human music.)

The ability of birds to distinguish rhythm was famously demonstrated by the scientist Irene Pepperberg's grey parrot Alex, who would move to a beat, and the domesticated cockatoo Snowflake, as analyzed by Aniruddh Patel of Tufts University, who moves to the rhythms of electronic dance music.

Carel ten Cate from the University of Leiden has studied the ability of parrots and songbirds to learn new rhythms and other musical features, and individuals within the same species appear to possess a broad range of abilities.

Why Birds Sing

Our guide in understanding why birds sing will be Ofer Tchernichovski, a professor at New York's Hunter College and author of the classic article "How a Zebra Finch Learns Its Song."

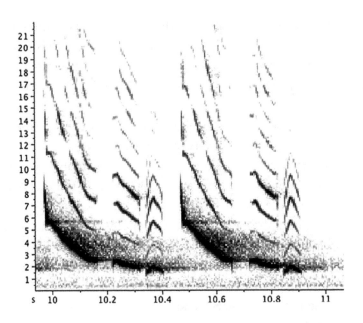

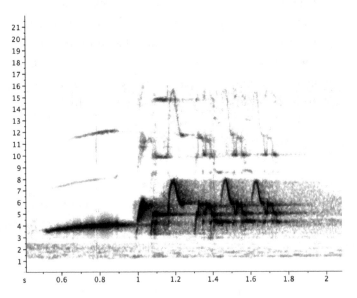

FIGURE 11.1 American songbird sonograms

Bird songs are usually analyzed as sonograms in which time in seconds is on the horizontal (x) axis and frequency in Hz on the vertical (y) axis.

The upper trace shows two downward sweeps during the song of a northern cardinal that each last about one half of a second. The fundamental frequency f_1 of these segments of this song starts at about 9.5 kHz (near D9) and over about 200 msec falls by nearly two octaves to about 2.5 kHz (about E♭7), followed by a gentler swoop and a short hump. The natural

Ofer says that birdsong not only attracts a mate and keeps competitors away but also promotes social bonds: indeed, he says that "at the population level, bird song is a culture."

The culture can be observed in duets between mates. In couples of black-faced rufous warblers, the males sing a "see," the female an "ooo," and the males an "eee," producing a "see-oooo-ee" song that sounds as if it is being sung by a single bird. Similarly, Karla Rivera-Caceres of the University of Miami found that canebrake wren couples in Costa Rica spend considerable time practicing to perfectly coordinate their duet into a seamless single song.

The paradigmatic studies of song culture by large flocks of birds are of white-crowned sparrows, whose range stretches from northern Alaska to Mexico City. The ornithologists Luis Baptista, Peter Marler, and Barbara De Wolfe found that white-crowned sparrows sing in local dialects, with songs of birds in a region similar to one another. Both sexes of white-crowned sparrows sing, with the female quieter and possessing a larger variety of repertoire.

During their first hundred days of life, the male white-crowned sparrows learn a song by imitating those of other local male birds, and they will sing that song throughout their life, which can last more than thirteen years. This approach helps the sparrows use the version of their song to identify their own regional flock, as if they hang specifically with fans of surf music or gangster rap or tango.

For domesticated songbirds, the most studied species are canaries, a species that has been selectively bred for centuries for color and singing ability, and zebra finches, social birds who live in large flocks in Australia and, in contrast to domesticated canaries, remain essentially identical to the wild birds.

---

harmonic series is quite strong up through f_5 or f_6. Playing such northern cardinal songs back at a lower speed, one suitable to our nervous system, can help us ponderous slow-thinking humans appreciate their beauty.

The lower trace shows a song by a wood thrush producing two notes sung at once. After one note at about 3.5 kHz is held for about half a second with prominent f_2 and f_3 harmonics, the other branch of the syrinx simultaneously sounds the sustained note. The peaks that run between about 5 and 8 kHz in combination with the held notes produce a gorgeous trill-like sound.

Source: Author.

Canary song learning has been explored in detail by Fernando Notte-bohm and colleagues at Rockefeller University. The males sing during the breeding season, stop when the season is over, and then relearn the ability to sing each year. Fernando's lab made the discovery that this occurs because of the annual death of old neurons and birth of new replacements in a brain region known as the higher vocal center (HVC), in the avian cortex. The birth and survival of the new neurons is highest during the fall after the breeding season, when adult males begin to learn their new song for the year. The birth of the new HVC neurons is triggered in part by testosterone, and injecting female canaries with testosterone can increase the size of these nuclei and induce them to sing more.

"People think that zebra finches have ugly songs, but they are wrong," says Ofer. One reason our species has such bad taste is that birds perceive, act, and presumably think faster than we can, experiencing humans as slow-moving bores. If we slow a finch song by 60 percent, which you might try, we begin to appreciate their melodies and how the individual bird's songs differ.

Zebra finch singing provides social bonds, and while they live in large flocks, they form monogamous pair bonds. The male learns his own song over the first three months of life by accurately imitating the complex songs of another zebra finch, typically his father, and learning to reproduce the original, in Ofer's words, "as fast as a brain can do: people cannot do it."

Female zebra finches perform unlearned innate calls of shorter dura-tion. Ofer's lab found that females have more precise rhythm than males and more rapidly learn to alternate short calls between individuals to pro-duce a "hocket" (see chapter 5).

Zebra finches will "work," that is, undergo an unpleasant experience, in order to hear songs sung by other zebra finches. A study by the late Kirill Tokarev from Hunter College reports that this likely occurs because dopa-mine neurotransmission is increased in males when they hear song. Male birds will "pay" to watch videos of other zebra finches sing by undergoing a mild punishment—a puff of air to the face. The study found that females will undergo mild punishment as well, but only to hear their own mates.

Bird song is thought to define membership in a group, something Ofer calls a "stable polymorphic culture," with some groups remaining stable and others changing the patterns of their song. The personality of each group can achieve a particular style by magnifying unique features of the songs. Perhaps this is like the way we identify the sound and performance of an individual musician, say, Miles Davis's trumpet from his trademark Harmon mute.

From these studies, we might guess that the purposes of birdsong are for practice, to declare territory and announce "I am here," to form bonds with a mate, and to define membership in a group.

Birds Who Imitate Other Species

In contrast to the songbirds who learn songs from members of their own species, parrots and parakeets can imitate human speech, as can members of the crow family, including ravens and even, occasionally, blue jays.

In the wild, the African grey parrot imitates other species, whereas Amazonian parrots of the Western hemisphere imitate only their own species and learn local dialects in the wild. As mentioned, the best-known individual parrot was Irene Pepperberg's African grey Alex, who learned to call objects by name.

Among the songbirds, the starlings, which include common starlings and myna birds, can sing both their own songs and imitate other bird species and other sounds.

In North America, our champion songbird imitator is the northern mockingbird. Both sexes sing their own songs and also mimic the sounds of frogs and other birds, including blackbirds, orioles, and aggressive species like jays and hawks. Mockingbirds continue to learn new sounds throughout their lives.

Another champion songbird imitator is the red-capped robin-chat of Africa, who has been heard singing the songs and calls of at least forty other species, including roosters and eagles, which they hear far above in the air. Thomas Struhsaker, affiliated with Duke University, reports that a single robin-chat in Uganda's Kibale forest can imitate the combined male/female duet of the black-faced rufous warbler *and* that two robin-chats can sing the warbler duet together, one having learned the male and the other the female part.

The world grand master vocal imitator in the wild is the lyrebird of Australia and Tasmania, the tail of which indeed resembles a Greek lyre, although lyrebird fossils have been dated to 15 million years ago, a bit earlier than Pythagoras. These large birds, nearly a meter long, live up to thirty years, and they sing and dance throughout the year, though mostly in the winter (June to August), during the mating season.

Both sexes sing their own song. Young lyrebirds learn the song of local adult males, and once the song is learned, it remains mostly unchanged. About 70 percent of their vocalizations, however, are calls, songs, and beak snaps by other bird species, and even the sounds of flocks of birds, as well as calls of wild mammals including koalas and dingoes.

David Attenborough's documentary *The Life of Birds* shows two lyrebirds who imitate a camera shutter, drills, hammers, and saws. Note that these are captive birds in a zoo and wildlife sanctuary. Wild lyrebirds have not yet been recorded who imitate human-produced mechanical power tool sounds.

The deservedly renowned example of wild lyrebirds imitating human-produced sounds is a population of "flute lyrebirds." A century ago, a family of potato farmers in the town of Allan's Water in the New England Tablelands of New South Wales are said to have raised a lyrebird as a pet in a house with a flute player. Whether the original bird was domesticated or wild, he taught other lyrebirds, and now generations of wild lyrebirds in the region sing a flutelike call including bits of melody from the Irish dance "The Keel Row" and a DO RE MI practice scale. Vicki Powys, Hollis Taylor, and Carol Probets wrote a detective-like study, "A Little Flute Music," to determine the veracity of this story and, as with all good scientific investigations, came up with additional questions. But do listen, as recordings of the flute lyrebirds are astonishing.

Night-Flight Calls

During the spring and fall, and nearly always hidden from us, a massive vocal choir of millions occurs during night-flight migrations. These enormous flocks are composed of species ranging from tiny sparrows to enormous herons, and some migrate thousands of miles. For example, that small virtuoso, the white-crowned sparrow, can fly over 300 miles in a single night during its annual 2,600-mile trips between Alaska and Southern California. Some black poll warblers, a tiny bird, migrate from Canada and Alaska to the Amazon, with the longest leg of their journey an average distance of 1,580 miles. This segment of their migration takes only three days, during which they fly without pause at twenty-five miles per hour.

The migrations typically occur at night so the birds can avoid predators like hawks and cats; because cooler nocturnal temperatures keep them from overheating and dehydrating; because the lack of daytime air thermals minimizes the turbulence they create and makes it easier to maintain a steady course; and because the moon and stars aid in navigation.

In contrast to complex songs, night-flight calls are short chirps or buzzes, typically a tenth of a second or less, and relatively quiet. The distinctive calls by a species or group of individuals probably help a flock to stay together while flying in the dark and may help them avoid midair crashes. Intense bouts of flight calling probably indicate severe instances of disorientation.

The paradigmatic study of the night-flight calls was conducted by the historian and ornithologist Orin Libby (1864–1952) in 1899. Libby listened from a quiet hill near Madison, Wisconsin, on a chilly September night with no cloud cover as the flocks flew south. He could identify many of the species from their flight calls. He counted 3,800 calls from the sky before he stopped at three in the morning. He estimated the number of birds migrating by counting those who crossed the face of the moon and multiplied that by the fraction of the sky that the moon occupied. From the 358 crossings he counted, he estimated that a minimum of nine thousand birds had flown over the hill that night.

The ornithologist Andrew Farnsworth from the Cornell Lab of Ornithology and colleagues analyze the patterns of night-flight calls using contemporary means to record and process the signals. One approach is to place directional microphones on skyscrapers pointing skyward and compare the recordings with live radar reports of the migrations from the National Oceanic and Atmospheric Administration. Current reports of bird migration, using methods to exclude meteorological phenomena routine in radar data, can be viewed on Cornell's BirdCast website. Their research has provided new information about the effects of exposure to light pollution on migrating birds, assessments of aircraft hazards from bird collisions, and a large-scale characterization of bird migration.

The National 9/11 Memorial and Museum's "Tribute in Light" commemorates the destruction of the Twin Towers of the World Trade Center by shining two intense beams of light skyward from where the towers stood. This occurs around September 11, a busy time for migrations. Andrew and his colleagues found that the light beams significantly disoriented migrating birds, causing them to gather, circle, fly more slowly, and call much more intensively. As a result of these findings and thanks to the work of volunteers from the New York City Audubon Society, the lights are shut off periodically when birds are in danger or gather in large numbers, allowing them to proceed.

Insect Music

There are something like 900,000 species of insects, but only a minority are thought to hear or make sound, at least as we define it. For example, of the 350,000 beetle species, only a small group of scarab and tiger beetles is known to have ears. Nevertheless, they all have organs that can sense vibration, so we might reconsider our current assumptions.

How Insects Hear

Insects use sound and hearing for familiar reasons: to attract mates, as a feature of aggressive behavior between males and to escape predators, and for some distinctively insect behaviors. For example, parasitic flies known as *tachinids* have ears placed on their necks that are sensitive to male cricket mating calls, allowing them to deposit their maggots on the crickets. Or some moths have ears near the base of the wings that detect very high frequencies in order to detect ultrasonic echolocation used by the bats who hunt them: when the moths hear the bats, they stop producing their own sounds, which they otherwise use to attract mates.

The first insects, which appeared about 400 million years ago, are thought to have been deaf, but all insects possess *chordotonal organs* that detect vibration and the relative movement between their body segments. For example, the chordotonal *subgenual* (meaning "below the knee") *organ* detects surface vibrations, as from a leaf on which an insect might perch. This organ is used to detect vibrations insects use to communicate with one another. Another type of chordotonal organ known as *Johnston's organ* is at the base of antennae and detects movements driven by wind or gravity.

Chordotonal organs are composed of multicellular units known as *scolopidia* that are typically found just under the exoskeleton. The scolopidia are in turn composed of sensory neurons and supporting cells. The sensory neurons are bipolar, with a single dendrite and axon. They perform a function analogous to the inner hair cells of the cochlea, with the dendrite possessing a primary cilium reminiscent of the mammalian hair cell stereocilium. As with hair cells, these neurons are stiff, and movement activates mechanosensitive ion channels that open very rapidly.

As observed in katydid fossils from 165 million years ago, some chordotonal organs evolved to detect sound. By tracing evolutionary pathways, Martin Göpfert from the University of Göttingen, a neuroscientist who specializes in insect hearing, estimates that chordotonal ears in insects evolved more than twenty separate times.

The chordotonal organs that detect sound are of two major types, a *tympanal organ* that is similar to the eardrum, and the aforementioned Johnston's organ.

Tympanal ears can be found in many parts of insect bodies, including a single *cyclopean ear* with two eardrums, found on the middle of the chest of the praying mantis. Some moths have ears within their mouths. Grasshoppers often possess tympanal organs on the first abdominal segment, but the bladder grasshopper *Bullacris membraciodes* possesses twelve ears on its abdomen.

The subgenual tympanal ears of katydids, crickets, and grasshoppers feature a hearing organ known as the *crista acustica* or *Siebold's organ*. In katydids, air waves enter tubes on the leg known as *trachea* to vibrate an eardrum (*tympanum*) from both the interior and exterior sides. The crista

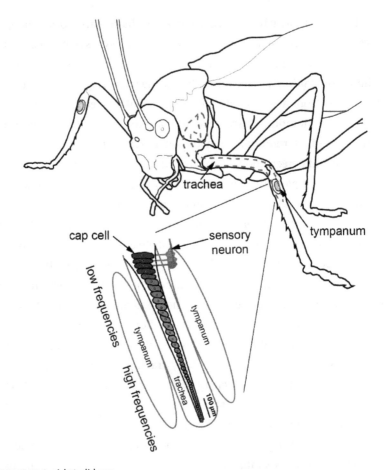

FIGURE 11.2   A katydid ear

The katydid's hearing organ, the *crista acustica*, is in the middle region of all six legs. Sound waves enter the trachea to vibrate eardrums from both the interior and exterior sides. Located alongside the trachea, the crista acustica detects sound-induced air motion that triggers mechanosensitive ion channels in sensory neurons. These channels are opened in the stiffer end of the organ by high frequencies and in the more flexible end in response to low frequencies, a tonotopic mapping similar to the mammalian ear.

Source: Manuela Nowotny (University of Jena). Used with permission.

acustica has a graded frequency response, with larger scolopidia at one end sensitive to low frequencies progressing to smaller scolopidia at the other end that vibrate at higher frequencies, in some cases as high as 300 kHz. It is striking that our distant relatives evolved to parallel the mammalian inner ear, with features analogous to the tonotopic organization of inner hair cells on the basilar membrane (figure 11.2).

In contrast to tympanal ears, Johnston's organ in the antenna is used to perceive sound by the *Hymenoptera*, the order that includes bees and ants, and by mosquitoes and fruit flies. The antennal ears respond to air particle movement that drive the antenna's outermost segment, known as the flagellum, to vibrate. In contrast to the ability of tympanal ears to sense very high frequencies, the scolopidia of Johnston's organ are usually excited by frequencies below 1 kHz.

The fruit fly *Drosophila melanogaster* is by far the most studied insect, mostly because they are useful for genetics thanks to their short reproduction time, producing the next generation of offspring in as little as twelve days. They also cost virtually nothing to grow, requiring only a jelly jar with a change of food every two weeks. Early genetic studies, about a century ago, were facilitated by enormous "polytene" chromosomes in the salivary glands of these flies, enabling gene mapping within the chromosomes. In the fruit fly, Johnston's organ contains about 480 scolopidia and is used to sense gravity, movement, and sound. The ability of fruit flies to detect sound is required for courtship songs and identifying the wing-beating frequency of mates. Sound is also involved in some aggressive behaviors between males.

In honeybees, Johnston's organ is used to detect sound from the buzzing of the wings around the range of middle C (C4) during the *waggle dance* that communicates the location of flower nectars to other bees. Their ear further appears to sense changes in electric fields caused by the dancing bee, a form of communication using a sense we lack and challenging for us to imagine.

The number of auditory neurons that run from insect ears to other regions of the nervous system is extremely variable, from only a single neuron in moth tympanal ears, which are specialized to hear bats, to 2200 neurons in cicadas and about 15,000 in the Johnston's organ in male mosquitoes, a number close to that in the human cochlea. The frequency of incoming sound appears to "phase lock" with neuronal firing, so that higher frequencies produce higher auditory nerve firing rates. The full auditory neuronal circuit can be very large in some flies, consisting of as many as 20,000 neurons, with the Johnston's organ nearly half of the diameter of the entire head.

The detection of sound by insects can be extremely rapid, as might be expected if one must constantly avoid bats to survive, much less a swat by a human. This is in part because of a very short auditory circuit that projects directly from the first synapse. This rapid listening can also be useful for mating; during katydid mating calls, females can respond to the male within 25 milliseconds.

How Insects Sing

The singing insects that we typically notice are male cicadas, katydids, grasshoppers, locusts, and crickets. Some other species, including moths, produce sounds that are too quiet or too high in frequency for our hearing.

The cicadas are the superstar insect vocalists, as only they are known to have developed a specific singing organ. Cicadas possess two curved plates known as *tymbals* in their abdomen. The abdominal muscles contract and relax the tymbals, and the abdominal chamber resonates like the body of a tiny cello to amplify specific frequencies.

Most cicada species sing in a buzzsaw-like whine—one species is known as the scissor-grinder cicada for a reason. Listening to the dusk-singing cicada will be a pleasure for fans of noise music. The song of Linnaeus's seventeen-year cicadas, which indeed spend seventeen years as nymphs underground to emerge on one night in May and die in July, is genuinely disturbing. The species ought to receive royalties for their use in horror movie soundtracks.

Cricket singing is far more popular with our species. Their songs tend to be center around 4 kHz frequencies, not far from our own musical preference. Nathaniel Hawthorne wrote in "The Canterbury Pilgrims": "He listened to that most ethereal of all sounds, the song of crickets, coming in full choir upon the wind and fancied that, if moonlight could be heard, it would sound just like that."

Crickets, along with grasshoppers and katydids, produce songs by rubbing the one forewing, the *scraper*, on the *file* forewing, a process known as *stridulation*. This generally produces high trilling sounds, although mole crickets produce frequencies so low that they sound like frogs. Tree crickets have evocative and simple trilled songs, and some, like the snow tree cricket, sing in synchrony with other males in outdoor cricket choirs. Cricket stridulation is faster when the air is warmer, and the pulse speed of the cricket songs can be used to estimate the temperature.

The common true katydid in the Eastern United States has a pleasant song of three syllables, which provide its name. Groups of katydids possess different regional dialects and, as seems appropriate, the southwestern katydid drawls in a lower pitch. According to the insect biologist Mark Moffett, who has studied insect life in the forest canopy: "It turns out that a katydid should be able to detect exactly where another katydid is located on the tree it's sitting on by how the vibration has been altered as it passes through different-sized branches or the trunk."

Locusts produce their sound differently: they are essentially violinists, rubbing the femur of their hind legs against the edge of their forewings like a bow on a string.

Even though most beetle species are currently considered as deaf, some create sound by rubbing a surface against an external surface with ridges, as if stridulating with another object.

Moving on to the *Hymeoptera*, the waggle dance of honeybees conveys information both by sound and other means of communication. Unacknowledged by us bulky obtuse humans, ants produce and are aware of sound at least under some circumstances. Mark Moffet writes: "Ants mostly employ scent for communication, but vibrations come up in the context of a 911-style emergency: Every time a human walks across a lawn, we crush the tunnel systems of ants underfoot. The buried ants emit a stridulatory squeak that informs rescue ants on the ground surface where to dig them up."

Female fruit flies require a highly discriminating auditory system: there are 1500 species, each with a unique song, and they are most receptive to the courtship songs of their own species. Male fruit flies produce their courtship song by vibrating one or both wings, to produce 100–400 Hz vibrations. We can hear these songs by placing a tiny microphone into their home in the lab, the fruit jar. The female hears these songs using Johnston's organ. The male's singing, together with dancing, licking, and tapping (tasting) the female with his forelegs, can lead to successful mating. The female chooses the male with whom she wants to mate primarily by assessing his song quality: if she doesn't want to mate, she kicks the male in the face until he leaves.

The neurotransmitter dopamine stimulates not only fruit fly courtship song. As noted by Aike Guo and colleagues at the Chinese Academy of Sciences, it also drives male-male courtship. These behaviors have been observed in both mutant flies with altered dopamine and after dosing flies with drugs that stimulate dopamine transmission, including aerosolized freebase cocaine and methamphetamine.

Some male moths sing by rubbing scales on their wings. Ryo Nakano from the University of Tokyo showed that the female Asian corn borer moth stops moving when she hears the male's courtship song. She also halts if she hears ultrasonic bat calls, suggesting that this species has adapted to sex a behavior that may have originally been in response to predators: playing dead.

In contrast to Asian core borers, the Japanese lichen moth reacts to bat calls by emitting ultrasonic clicks to jam the bat's sonar.

Singing by Other Primates

My vote for the land animal with the most astonishing song is the gibbon. These are highly endangered apes encompassing nineteen species in Southeast Asia. Their song bouts last from ten minutes to a half-hour in the morning and are loud enough to be heard more than a mile away. Some of their singing sounds startlingly like whales.

Both sexes sing, with females producing a "great call" that lasts about twenty seconds. They are monogamous, and the pairs sing in duets to advertise or reinforce their bond. Single gibbons sing to attract mates.

Esther Clarke and collaborators at the University of St. Andrews analyzed gibbon songs at Khao Yai National Park in Thailand. They built full-size models of leopards, tigers, pythons, and eagles, using catapults to hang them over branches of trees. They concluded that gibbons also use songs to repel intruders, using the same notes as in their mating repertoire but assembling them differently.

Lemurs, also fellow primates, are native only to the island of Madagascar. Of one hundred or so species, all are under threat of extinction. The indri is the largest and is known for "contagious" calling: when one starts a loud call, the rest of the community two years of age and older joins in.

Both male and female indri sing. They can perform in duets, like gibbons, and with glissandos that resemble humpback whales. Marco Gamba from the University of Turin in his paper "The Indris Have Got Rhythm!" analyzed singing in indri social groups dominated by females, who typically have a monogamous relationship with a male. The dominant female and male in a group seem to sing together more than the nondominant, nonpaired members. The authors suggest that the nonoverlapping phrases might "advertise" individuality and potential availability.

Howler monkeys, the largest New World monkeys, live in southern Mexico and through much of South America. Their calls, generally at dawn and dusk, can be heard three miles away. The songs can be individual solos or group performances. Both sexes call; the females are quieter and overlap more. Margarita Briseño Jaramillo and colleagues from the National Autonomous University of Mexico studied calls of black howlers in Palenque National Park. She identified twelve different calls, including roars, barks, moos, and metallic cackling: you can hear singing that will provide inspiration for heavy metal bands. The moo was only used when monkeys were reunited after separation. The grunt is used during play. The bark is used during conflicts with northern brown howler monkeys.

Sound Under Water

The French oceanographer Jacques Cousteau produced a film called *Le monde du silence* (The silent world), but the whale song pioneer Roger Payne calls the ocean "a very loud place." This is in large part because of propeller ships, oil drilling, and sonar. For example, the ambient noise at 30–50 Hz frequencies measured west of San Nicolas Island in California was 10–12 decibels (dB) higher in 2004 than 1964, a period that saw a fourfold increase in shipping.

In chapter 1, we discussed how decibels report volume relative to a constant value of "excellent hearing" set at an air pressure of 20 µPa. Liquids are far less compressible than gases (at 200 times normal air pressure, water volume is decreased by 1 percent), but sound produced in water still forms pressure waves.

The volume of sound in water is similarly reported in dB, but by convention it is set relative to a water pressure standard of 1 µPa. This is a tiny amount of pressure: water pressure at sea level is already at about 101 kPa, and it climbs another 10 kPa with each 10 m of depth. The different reference pressures in water and air work out to an equivalent sound pressure reported as 26 dB higher in water than air.

MATH BOX 11.1

An extremely loud humpback whale has been reported as 176 dB "loud," which corresponds to

$$176 \text{ dB} - 26 \text{ dB} = 150 \text{ dB in air.}$$

To find the increase in water pressure by this very loud humpback, we can rearrange the equation for dB in air in chapter 1 for water as

$$176 \text{ dB} = 20 \times \log_{10} (x \text{ µPa}),$$

$$176 \text{ dB}/20 = 8.8 \text{ dB} = \log_{10} (x \text{ µPa}).$$

To solve for $\log_{10} x$, raise both sides of the equation by an exponential of 10.

$$x \text{ µPa} = 10^{8.8} \text{ dB} = 630{,}957{,}344 \text{ µPa} = 0.631 \text{ kPa.}$$

So even a sound that is immensely loud changes water pressure by a tiny amount. This extremely loud 176 dB song at a 100 m depth (201 kPa) increases the pressure by only

$$0.631 \text{ kPa} / 201 \text{ kPa} = 0.3\%.$$

Ocean sound has effects on animals that can't "hear," at least by means we now understand. For example, coral larvae settle in areas with relatively high levels (10 dB) of low-to-mid-frequency (24–1000 Hz) sound, which appears to reflect a healthy reef. Fish similarly appear to hear as they use sound to find suitable habitats.

As you might suspect, fish hearing is surprisingly understudied, and while most species are thought to be deaf, the sand lance hears between 50 and 400 Hz and responds to the sounds of its major predator, the humpback whale.

But some fish produce calls and so presumably must also hear. These include freshwater and seawater drums and ocean clownfish, toadfish (the most studied vocal fish, who hums at night), "grunters," and tiger perch. Similar to songbirds, many of these species tend to call together in a "dawn chorus" and at times related to the phase of the moon.

Sea invertebrates such as squid and crab hear using organs known as *statocysts* that also sense balance and acceleration. These are two fluid-filled sac-like organs near the base of the brain lined with hair cells; they resemble ears. They contain a few grains of sand or a small mass of calcium carbonate known as a *statolith*. In response to sound, the hair cells activate the statolith (rather than the organ of Corti) to generate signals sent to the nervous system.

The range of frequency perception a sea invertebrate can hear varies by species. Hearing has been detected in crayfish from 20 to 2350 Hz and by prawns between 100 and 3000 Hz. According to T. Aran Mooney from the Woods Hole Oceanographic Institute in Massachusetts, longfin squid can hear frequencies between 30 and 500 Hz, a range well adapted to perceive waves and wind but not useful for detecting the echolocation signals emitted by their major predators, the toothed whales, including dolphins.

Sea invertebrates respond to changes in sound volume. Mooney reports that when he played sufficiently loud sounds, squid released ink, jet propelled themselves, and translocated pigment within specialized cells known as *chromatophores* to change their color and alter their body pattern, a response typically used for camouflage. Loud human noises such as sonar can damage the statocysts and thus kill squid.

Singing Amphibians and Reptiles

The calls of male frogs and toads are, one hopes, familiar to the reader. They are involved in mating and chorusing and display regional dialects. Their nervous system is exquisitely attuned to the presence of signals from their potential mates, including in species in which the female taps on surfaces to advertise her readiness.

The biological underpinnings of these vocalizations have been extensively studied by Darcy Kelley and colleagues at Columbia. They examined the African clawed frog—the genera *Xenopus*—who sing underwater during the breeding season. Using hydrophones (a microphone used to record sound underwater), they found that the male's song's sounds are two-note pairs, with harmonic intervals shared by related species.

They also found that the larynx in the *Xenopus* species is tuned to produce different sound pulse frequency intervals. These pulses are created by the larynx, which can be induced to sing even when removed from the frog, provided that the vocal nerves are stimulated in the species-specific pattern. A species' specific song can be evoked directly from the brain after it is removed and exposed to serotonin, allowing *Xenopus* researchers to uncover the neurons responsible for the evolution of different song patterns. Females are specifically attuned to the pitches and harmonies of their own species' males, suggesting that the production and perception of songs evolved together.

In 1923, D. H. Lawrence composed the poem "Tortoise Shout," describing vocalizations during tortoise sex.

> And giving that fragile yell, that scream,
> Super-audible,
> From his pink, cleft, old-man's mouth,
> Giving up the ghost,
> Or screaming in Pentecost, receiving the ghost.

Setting aside Lawrence's apparent familiarity with their shout, our species did not realize until very recently that turtles vocalize and hear. Camila Ferrara, Richard Vogt, and Renata Sousa-Lima from the National Institute of

Amazonian Research reported in 2013 that not only do female giant Amazon River turtles produce eleven types of sound, mostly but not only in the river, but that hatchlings also produce sound even while inside the egg.

These vocalizations are involved in gathering adults and hatchlings together in mass migrations; they also provide the signal by which hundreds of turtles emerge from the sand of a beach in the Amazon within a span of minutes to lie in the sun. The authors suspect that all turtle species vocalize.

Why was this previously missed? In part because the sounds are made mostly underwater and because, as you might guess, turtle songs are slow, not often performed, and use low frequency sounds, so that one needs to record far from human noise pollution. A sadder reason is that in captivity and zoos, they simply stop singing.

Above the water, the reptile most studied for call production is the tokay gecko, although multiple lizards hiss and growl. Carl Gans and Paul Maderson of the University of Michigan and Brooklyn College classified three overall types of sound production in modern reptiles, including the hisses and tail rattles of snakes and the calls of crocodiles.

Whale Song, Calls, and Hearing

The discovery of whale song seems similar to Columbus's discovery of the New World, where people had been living for at least 15,000 years.

An early report of whale song is attributed to a whaler, Captain William H. Kelly, who in 1881 heard a struck whale groaning as he put his ear to a harpoon line in the Japan Sea. According to Kelly's colleague, Captain Herbert Lincoln Aldrich, writing in 1889:

> It has been known for a long time that humpback-whales, blackfish [the pilot whale], devil-fish [the gray whale] and other species of whales sing, and that walruses and seals bark under water, and it is believed that all animals having lungs and living in the water, as these do, have their own peculiar cry, or as whalemen express it "sing" . . .
>
> With bowhead-whales the cry is something like the hoo-oo-oo of the hoot-owl, although longer drawn out, and more of a huing sound than a

hoot. Beginning on F, the tone may rise to G, A, B, and sometimes to C before slanting back to F again. With the humpbacked-whale, the tone is much finer, often sounding like the E string of a violin.

Captain Aldrich was correct, as all eighty-six of the currently designated species of whales, which include dolphins and porpoises, vocalize, ranging from grinding clicks produced by narwhals to the well-known squeal of the porpoise and the long and extremely complex songs of humpbacks.

There are two families of whale species, and they vocalize differently. The majority of whale species are toothed predators (*odontocetes*), including dolphins, narwhals, and orca, who, in addition to other calls, use echolocation to image their environment with ultrasonic signals.

Fourteen whale species are baleen whales (*mysticetes*), named for their baleen plate, which sifts seawater while the whale swallows large amounts of minuscule prey. These include humpbacks, bowhead, blue, fin, right, and minke whales. The baleens are master singers with a complex repertoire.

Whales are the only mammals that have evolved ears adapted for underwater hearing, and they perceive the broadest frequency range of any animal known. During their long evolution from land mammals, the ears of whales lost their pinnae and outer ear and migrated away from the skull to behind the base of the jaw. While an ear canal still exists, it is not directly connected to the eardrum and is not believed to conduct sound. Instead, whales hear by picking up the vibrations from the water that travel through the head, including through blubber, to the jaw and mandibles, which conduct the signal to the ear. Their middle ears are protected from the large pressure changes between breathing air and deep dives by dense bony ossicles and eardrums.

A comparative anatomical study by Darlene Ketten from Harvard Medical School reports that the ears of toothed whales are adapted to sensing frequencies above 150 kHz, that is, the high range of bats, while baleen whales appear to have more acute hearing for low frequencies. The whale ear's basilar membranes cover wider frequencies than those of terrestrial mammals and are associated with a far greater density of nerve cells, especially for toothed whales, giving them an outstanding ability to distinguish frequency, an important feature for echolocation.

It is not straightforward to assess the sensitivity of whale hearing for long-distance communication. Beluga whales in Bristol Bay, Alaska, can hear at volumes as low as 35 dB in water (corresponding to the pressure of 9 dB in air: for comparison, audiologists rate hearing sensitivity of 20 dB to be normal for us). They also hear a far greater range of frequencies than we do, on average from 22 to 110 kHz, with some individual beluga whales found to hear frequencies as low as 4 Hz and as high as 150 kHz: two octaves lower and three octaves higher than we do!

This provides belugas with a sophistication and detail for perceiving the sound of their environment we can only imagine. One might assume that as for other animals, including us, they are particularly tuned to sounds produced by their own relatives. Whales may participate in a massive orchestra over great distances, using song to maintain their awareness and to harmonize with one another.

Songs of the Baleen Whales

Baleen whales evolved not only ears for underwater hearing but also an extraordinarily expressive voice for singing. Joy Reidenberg from the Mount Sinai School of Medicine and colleagues find that the baleen whales contract muscles in the throat and chest that drive air flow between the lungs and an inflatable sac in the larynx next to structures similar to the vocal cords, the *u-folds*, causing them to vibrate and produce sound. This vibration propagates through the surrounding tissue into the water. The change in frequency and volume of the voice can be controlled by the shape of the sac in the larynx. This mechanism of singing doesn't expel air, allowing the whales to produce very loud sounds without the need to come to the surface to breathe.

One of the singing whale species heard by Captain Kelly, the bowhead, spends its life entirely in Arctic waters. It can grow to fifty-nine feet in length and weigh one hundred tons, with the largest mouth of any animal. While formerly driven close to extinction by whaling, the population has recovered in Alaska since commercial harvesting has been halted. Bowheads are named for an arch on their head that is used to punch breathing holes in ice sheets. They sing with an immense variety of different sounds. Kate Stafford from the University of Washington and colleagues recorded 184 different songs from bowheads over three years off the coast of Greenland during the winters, under nearly complete ice cover and total darkness.

The humpback whale, at up to fifty-two feet long and thirty tons, lives in every ocean and is by far the most studied and widely appreciated cetacean singer. When the world population decreased to five thousand animals in 1966, the brink of extinction, the International Whaling Commission banned commercial humpback hunting, and the current worldwide population is now estimated to have risen to eighty thousand.

In striking contrast to bowheads, who sing under the Arctic ice, humpbacks migrate, spending the summer months near the poles, where food is plentiful, moving to tropical oceans to breed and sing during the winter. Some travel over five thousand miles from Antarctica to Costa Rica. Humpbacks that spend the warmer months near Iceland and Norway arrive at breeding grounds off the Dominican Republic and Puerto Rico at the end of February, to be joined a bit later by whales migrating from the eastern coast of North America. One highly recorded population spends most of the year off the coast of Alaska to return each year to breeding grounds in Hawaii. Louis Herman from the University of Hawaii has recorded individual males singing for over twenty years, a span that allows us to hear how their songs change over a good proportion of their life.

Our contemporary understanding of whale song is thanks to efforts of the American biologist Roger Payne and the poet-explorer Scott McVay, who heard recordings made with a hydrophone by Frank Watlington, a navy engineer, who accidentally taped whales off the coast of Bermuda while listening for Russian submarines. Reminiscent of Captain Kelly a century before, on one exceptionally quiet evening, Payne heard a humpback singing through the bottom of the boat.

Payne and McVay wrote that the humpbacks produced songs that lasted for seven to thirty minutes and that they repeated the songs precisely, with each individual whale adhering to his own song. The songs mostly use frequencies we hear (mostly 8 Hz to 10 kHz) and are very loud (from 151 to 173 dB underwater). They wrote that the principal difference between songbird and humpback songs is that bird songs usually last for a few seconds, while humpback songs last for minutes. They point out that one bird song is usually separated from the next by a period of silence but that humpback songs are repeated without a significant pause or break in rhythm.

Roger released the record album *Songs of the Humpback Whale* in 1970, a bestseller that popularized whale conservation and the environmental movement and helped bring about the moratorium on commercial whaling by the International Whaling Commission in 1982. The album features both solo and group songs and contains three recordings attributed to Frank and two by Roger and Katie Payne, whose work on subsonic sounds by elephants we will come to. They and others have also produced commercial recordings of blue and right whale songs.

I think that the recordings of whale music have helped save them from extinction by our species.

Humpback Whale Conservatori

Only male humpbacks sing entire songs, both alone and in groups; females make social vocalizations during the summer feeding months. Both sexes produce percussive sounds from flipper or tail slaps. For the males, singing is very common in the warm-water breeding grounds in the Northern Hemisphere during the winter, increasing between mid-February and mid-March, a period coinciding with ovulation. Singing has also been recorded during migration and occasionally in the subpolar feeding grounds in late spring and late autumn, possibly by younger males who are practicing.

During the winter, in addition to solo songs, male choruses can be heard singing different sections of the song simultaneously, like a canon. This chorus is speculated to be a *lek* that females without calves visit for mating. Leks are formed during mating season not only by whales but by sea lions, harbor seals, walruses, and dugongs, all of which also vocalize throughout the breeding season. It seems that females don't visit lone singers but often leave the lek pregnant.

Male humpbacks learn songs both from their father and unrelated males, and individuals have their own song but can change it or learn a new one. A population within a shared ocean basin usually conforms to a dialect or song type. Sometimes humpback songs change within the population over the years, but Michael Noad and collaborators discovered occasional "revolutions" when a complete complex new song takes over a whole population in about a year.

Roger Payne suggested that the new songs arrive from individuals moving from one breeding population to another or during a shared feeding ground or migration. Similar new songs have been recorded in extremely distant locations, such as both the Caribbean and Cape Verde Islands off the West African coast, or in both Mexico and Hawaii, or in both Eastern Australia and French Polynesia, or in both northeastern Brazil and Gabon in West Africa, and even across the African continent on both the Atlantic and Indian Ocean coasts. This suggests that new songs are learned in summer grounds and that by winter most singers have converged on the new style.

Long-Distance Low-Frequency Calls of Fin Whales and Elephants

Toothed whales produce calls in part to notify others of their presence. The most studied long-distance call is from fin whales, which are found in deep water on all sides of pack-ice fields that often extend for hundreds of miles, particularly in the Antarctic during spring. Hearing another fin may help them find their way through the ice packs.

Like elephants, fin whales use low-frequency sounds, around 20 Hz, apparently to trigger their location as the herd moves. These "blips" of sound can be very loud trains of near sine wave pulses or can be pairs of pulses that last about one second and are repeated at regular intervals, about five times per minute for fifteen minutes. They are followed by a period of silence of about two-and-a-half minutes, suggesting that the calls are interrupted by breathing.

Roger Payne and Douglas Webb from the Woods Hole Oceanographic Institute estimated the distance that fin whale vocals would travel based on the background noise of the sea, which is far louder now than before the advent of propeller ships and sonar. As with the bass frequencies entering your apartment from your neighbor's party, low frequencies, particularly around 20 Hz, travel well in water under ice, as shorter wavelengths are reflected by the rough surfaces of the ice. They estimated that before steamships, fins might detect another fin from 450 miles away over an area of 610,000 square miles. If their hearing were somewhat better still, they may be able to hear another fin whale from across the Arctic Ocean. The fin whale orchestra may connect extremely far-flung social groups scattered across the oceans of pitch-black, impenetrable ice fields.

Elephants also use low frequencies for long-distance communication. They are highly vocal and use a wide vocabulary of sounds for communication, including snorts, grunts, squeals (usually for delight, in my experience), and the well-known trumpet. Part of their diversity of sounds is because they use their trunks to produce some sounds and their mouths for others.

The production by elephants of very low frequencies has been described in anatomical studies of the larynx by Christian Herbst and colleagues from the University of Vienna. As you would suspect, elephants have the largest mammalian larynx known, and they are thought to make calls down to 10 Hz. Herbst concludes that some of the sounds are made by vibrations of the vestibular folds, which our species uses in Tuvan "throat singing" and death metal growls.

These infrasonic sounds were discovered by the biologist Katie Payne, who at the Portland Zoo felt the elephants produce a very low-frequency rumble, under 20 Hz, thus inaudible to us. (She felt the sound as a vibration.) To demonstrate their existence as sound, she simply sped up her tape recorder's playback speed.

The rumbles shake the dense molecules of the ground and propagate long distances. Payne and colleagues estimate that low-frequency rumbles are probably heard in East Africa for about 50 square kilometers surrounding the elephant and under some atmospheric conditions perhaps even up to 300 square kilometers. Elephants eat a tremendous amount of vegetation (Asian elephants eat about 400 pounds per day) and need to spread across a large area. The low-frequency sounds are used during their travels to keep the pack together, and they are now used by scientists in the Elephant Listening Project to track and study elephant herds in Africa.

Other animals that produce infrasound include the rhinoceros and the giraffe. Elizabeth von Muggenthaler and colleagues from the Fauna Communications Institute in North Carolina have described low-frequency singing by the rare Sumatran rhinoceros in the Cincinnati Zoo, which sounds similar to whale song.

Surprisingly, the far smaller koala possesses a vocal organ that allows them to produce bellows as low as 9 Hz.

Alligators produce underwater infrasound, using rumbles to produce waves in the water.

Echolocation by Bats and Toothed Whales

One cannot imagine two groups of mammals more different than bats and whales, yet both sing—singing by male bats has been relatively little studied but can sound very much like birdsong.

Bats and toothed whales tend to hunt in the dark, and both have developed a very different sense of sound, *echolocation*, in which they emit high-pitched clicking sounds that bounce back from prey: the shorter the time for the return, the closer the object. Moreover, because of the Doppler effect discussed in chapter 1, as the distance to the reflective surface (the prey) decreases, the echo's frequency increases.

Echolocation was discovered and named by Donald Griffin, an undergraduate at Harvard working with the physicist G. W. Pierce, who developed a means to detect ultrasonic sound. Griffin and a fellow student, Robert Galambos, proved the existence of echolocation to skeptical observers by showing first how bats avoid flying into wires suspended from a ceiling but collide with the wires when their mouths are tied shut or ears plugged.

Bats mostly generate the clicks in the larynx. These calls are short, from 0.2 to 100 milliseconds, and they range in frequency from 11,000 Hz, within our hearing range, to over 200,000 Hz, more than three octaves above pitches we perceive. The volume ranges from 60 to an ear-shattering 140 dB, particularly in open skies. Bats exhale and produce clicks during the upstroke of their wing. Some species decrease the volume as they get nearer to reflective surfaces to prevent damaging their own ears.

When searching for an insect, bats typically click at about twenty times a second but increase the pulse rate as they hone in, sometimes to as high as two hundred clicks per second. In addition to distance and movement, bats can estimate the elevation of their targets from the echoes reflecting from their *tragus*, a flap of skin within the ear. The inner ear is further specially constructed, with the cochlea longer in regions corresponding to typical call frequencies.

The auditory cortex of the bat has been studied by Nobuo Suga and colleagues at Washington University at St. Louis. A region they call the *FM-FM area* responds to the call and echo, with neurons responding to a specific time delay, thus reporting the distance of the bat to the target. Another known as the *CF-CF area* detects Doppler shifts and thus the change in distance to the target.

Toothed whales differ from baleens in how they control their voices. The beluga or white whale vocalizes by forcing pressure through its nasal cavities, which have a set of lips. Their clicks are of a wide variety, including the very low 20 Hz long-distance calls of the fin whale.

Toothed whales possess the ability to lower their pitch into the human range. Sam Ridgway from the National Marine Mammal Foundation and colleagues recorded a beluga whale named Noc (*no see*), who was trained for operations with the U.S. Navy, mimicking human speech, including ordering a diver, Miles Bragget, "out" of the water. Keepers at the Vancouver Aquarium report that another beluga whale would speak his own name, "Lagosi." Elena Panova and Alexander Agafonov from the Russian Academy of Sciences report that a captive beluga whale in an aquarium in Crimea who was housed with bottlenose dolphins learned to imitate the signature whistle calls of his dolphin tankmates.

It appears that all toothed whales, including dolphins, porpoises, beluga, orca, and sperm whales, use echolocation. Dolphins use very short clicks of about 50 microseconds at frequencies as high as 150 kHz. The clicks are emitted as a beam in the direction that their head points to plus a small incline upward. Different click frequencies produce the well-known barks and squeals of the bottlenose dolphin.

Recall that sound travels over four times faster in water than air. As with bat echolocation, the rate of toothed whale clicks can be altered as the target is closer, so that the estimate of distance is not confused by overlapping echoes. The echo is received similarly to the baleen whales: in the lower jaw and conducted to the middle ear. The whales can also deduce the type of object from the echo.

Toothed whale clicks can be immensely loud, up to 230 dB underwater for the Atlantic spotted dolphin. The distance that echolocation can be used to detect a small object in experimental settings is about 100 meters. Sadly, the entrapment of scarred whales and dolphins in fishing nets and toothed whale deaths from boating accidents indicates that their echolocation can be confused by our technology.

Animals Improvising on Musical Instruments

Here's a syntactically simple sentence fraught with semantic complexity: Can other animals play musical instruments? To paraphrase President Bill Clinton, it depends on what you mean by the words "can," "other," "animals," "play," "musical," and "instruments."

Circuses have long trained seals and walruses to use their snouts to play tuned car horns in a sequence, and street performers formerly trained monkeys to use sticks and play drums. But will nonhuman animals use instruments to create their own music?

Chimpanzees and Bonobos

The primatologists Jane Goodall and Adam Arcadi of Hofstra University studied drumming by apes in the wild. Chimpanzee drumming is used to communicate between distant individuals who are not in visual contact. In the forest, chimpanzees use their hands and feet to hit trees and play low-frequency drumming sounds that can be heard by humans a kilometer away. In the Tai National Park of the Ivory Coast, the chimpanzees usually drum while calling in "pant hoots," whereas in Kibale National Park in Uganda they tend to drum on trees without calling. The Tai chimpanzees drummed in distinct styles, while those in Kibale are reported to drum in a collective style. Some naturalists, including Peter Marler, suspect that chimpanzee drumming is similar to the well-known chest beating by mountain gorillas.

Valérie Dufour from the University of Strasbourg and colleagues studied spontaneous music making on human-made drums by captive chimpanzees living in a research center in the Netherlands. The longest-duration performance they observed was by Barney, who in January 2005 spontaneously performed on an upturned bucket in a style reminiscent of a bongo drum in a solo for over four minutes.

Together with the late Gordon Shaw, I explored whether bonobos, the close relatives of chimpanzees, would play on human-designed instruments. Gordon was a physicist at the University of California at Irvine who developed the notion of the "Mozart effect," his theory that children would become smarter if they listened to and learned to play Mozart.

We brought tuned musical bells to the bonobo colony at the San Diego Zoo. They immediately played the bells but after a few minutes realized it was better to throw and smash them against the walls. Bonobos are far stronger than any human; smashing the bells required no effort.

Bonobo colonies, both wild and in the zoo, have a matriarchal culture, and when I played a marimba we designed for them, they ignored it at first, until Lena, the matriarch, cuddled up to my leg for perhaps an hour. After that, the others began to pay attention. They would listen and react, but I didn't observe them playing the marimba during the few sessions we had with them.

Far more impressive results were achieved by the musician Peter Gabriel, who spent many hours singing and playing electronic keyboards with two bonobos, Kanzi and Panbanisha (figure 11.3), as well as from the efforts of the primatologist Itai Roffman from Yezreel Valley College in Israel, who has been working with Kanzi.

The psychologist and primatologist Susan Savage-Rumbaugh has lived and worked with bonobos for decades, beginning with Matata, who was imported to the Yerkes primate center in Georgia at puberty, after having learned the culture of wild bonobos. Matata's son, Kanzi, grew up with his sister Panbanisha and Matata in a language laboratory with a two-hundred-acre forest in Georgia in the 1990s, alongside Susan's own son. Kanzi understood spoken English, and with coaching learned to make stone tools, start

FIGURE 11.3 Kanzi playing a synthesizer keyboard
Kanzi improvising on a synthesizer during a session with Peter Gabriel, also on a synthesizer, behind the enclosure, with Susan Savage-Rumbaugh (right) coaching.
Source: Susan Savage-Rumbaugh. Used with permission.

fires, and paint. He also learned to speak both Matata's language and could speak to humans by manipulating geometric symbols to communicate his wishes and feelings.

In Susan's words:

> The bonobo language is quite musical, and often at night bonobos sing together before going to sleep. One of many "bonobo rules" set by their matriarch is that no one should go to sleep until everyone is happy with everyone else, which they express by singing with considerable jubilation each night.

Reading about Kanzi and his sister, Panbanisha, the musician Peter Gabriel wondered if they might be musically inclined and requested a visit. Peter brought his electronic keyboard and played to them, and they were offered a similar keyboard and urged to play along. They were polite and pressed a few notes but mostly listened to Peter. He continued to play for over a week, several hours per day, and they continued to listen but not "play music."

One afternoon when they were again listening to Peter but not playing, I noticed that they were also listening to Matata, who was about five hundred yards away. I decided that since they were paying attention to her and sending sounds her way, that I could request that "they play a song" to their mother. They immediately brightened, and I said to Peter, why don't you all play a "Matata song." Peter began, and this time they played keyboards with him and also sang to Matata.

It seemed that we found a key. Music needed a functional purpose, a topic. To co-create with Peter, they needed to make music together. They played songs about apples, bananas, grooming, and other bonobos. Peter immediately emulated their rhythms, chords, and melodies, and they took note. Prior to this, human musicians had always expected them to do what they did, but while bonobos could follow it, they did not find it interesting. Here was a human playing with them the kind of music they played, music that had a purpose. They tuned in. Peter felt that they had a "sense" of music already available to them.

About fifteen hours of video data were created by Peter Gabriel and the bonobos. Kanzi was more interested in rhythm and Panbanisha more in melody. It was clear that their songs have a beginning, middle, and end, as does a conversation. Watching the videos reveals that while Kanzi and Panbanisha are not remotely as skilled as Peter, they are making music with him of their own free will, not for a grape, and that they enjoy it.

It might be thought that Peter was simply "making the bonobos sound good" rather than co-creating songs with them, but our analysis reveals that this is not what took place. In Panbanisha's "Grooming Song," in which she played 226 notes, they go back and forth, with each taking turns as the leader or the follower, thereby co-creating the song. At one point, she wanted to alternate between the highest and lowest notes of an octave and signaled this to Peter with her fingers before she does so. He saw her intent, and they both alternated the octaves. To do this requires a sense of what an octave is, a knowledge of the location of the notes on the keyboard, an understanding that the lead can alternate between partners, all while continuing the song and maintaining the rhythm. None of these things were taught to Panbanisha, and prior to Peter's visit no one had ever co-played or co-created music with her. She had been allowed to explore the keyboard, and any knowledge she accrued of the location of various keys and notes she taught herself.

Altogether, studies by Jane Goodall and others in the wild and of apes in captivity demonstrate that that chimpanzees and bonobos have behaviors that we generally claim exclusively for ourselves.

Reminiscent of how Roger Payne and Scott McVay's humpback recordings helped the protection of whales, Itai Roffman, Susan Savage-Rumbaugh, and other students of ape behavior wrote a manifesto in 2019 calling for chimpanzees and bonobos, our closest relatives by genetic analysis, to be classified as hominids with the same official and legal status as our species. One hopes that this well-reasoned legal reclassification will help preserve and protect the chimpanzee and bonobo societies and cultures in the wild.

Songbird Instrumentalists

While one might guess that performing on musical instruments would be particularly successful for vocal imitators like mockingbirds, only limited efforts have been made for developing ergonomic songbird musical instruments.

The French artist Céleste Boursier-Mougenot has developed a means for zebra finches to make music in art galleries, by allowing them, for example, to alight on and thus trigger active electric guitars. In Austria, Reinhard Gupfinger and Martin Kaltenbrunner have built some creative and ergonomic musical instruments for captive grey parrots.

For his zebra finch colony, Ofer Tchernichovski rigged a lever that they could peck to hear the singing of another zebra finch, which in some cases they will do hundreds of times a day and even hundreds of times an hour.

Douglas Repetto from Columbia University, Linda Wilbrecht from Rockefeller University, and I extended that approach using a set of multiple levers to offer a means for zebra finches to trigger songs by other birds or other sounds. We were amazed to find that for some human music, including from trumpets that sound a bit like zebra finch calls, they would press the levers hundreds of times a day, although there was no reward other than hearing the sound. Very often, they would vocalize at the same time that they triggered the instruments.

The zebra finches had a musical "taste": some sounds, including punk rock guitars, would be played only once and not returned to. In addition to human-made sounds, they would press for some bird calls from other species: once they triggered a canary call, a species that can attack the smaller zebra finches, and, as apparent from their loud squawks in response, the canary call frightened them. They did not trigger that lever again.

No one has yet tested this kind of bird instrument in the wild, although Ofer suspects that they would play an instrument if it were ergonomically designed and placed. Until this is attempted, we won't know, to paraphrase Maya Angelou, why the caged bird plays.

I would take Ofer's side of the bet for some species: consider that the cedar waxwing has two common calls but is considered to have no full song, whereas a male brown thrasher can have a repertoire of over one thousand songs. I guess that an ergonomically designed instrument in the wild might be played at times by a male brown thrasher and also by his favorite intended audience, the female brown thrasher.

The Thai Elephant Orchestra

Of all the other species that we have learned to domesticate, the Asian elephant is the one for which future survival is in doubt. With rapidly falling numbers of both wild and domesticated animals caused by loss of habitat and less need of their traditional work roles with humans, their best long-range hope is now in the establishment of sufficiently large breeding populations in managed semiwild habitats. For the foreseeable future, these will have to be funded by a mixture of tourism and government and private support.

Domesticated for thousands of years but genetically no different from the wild populations from which they were captured, the Asian elephant has been trained in skills that in breadth, variety, and complexity certainly surpass any other animal. Formerly used in war as conveyors like horses and as effective weapons and later as loggers and commercial transport, their most recently developed employments are in the traditional human arts, including painting, musical performance, and soccer. These new roles have been invented over the past three decades, and they at least begin a new chapter in the long history of elephant training, one we hope will be safer and more pleasant than past eras.

By far the largest attempt at instrumental composition by other animals was the Thai Elephant Orchestra, based at the Thai Elephant Conservation Center (TECC), which is run by a branch of the Thai government, the Forest Industry Organization (FIO), and located between Chiang Mai and Lampang in northern Thailand. The TECC is a government-owned center that was the first devoted to the care and conservation of out-of-work domesticated elephants. There are now about one hundred privately owned centers as well.

The TECC was initially a home for logging elephants that belonged to FIO but were no longer employed, after a nationwide logging ban instituted in an attempt to reverse the country's deforestation. The center's elephant population, which typically hovers around sixty, is supplemented by other working elephants that owners can no longer maintain, donations from zoos, and "white elephants" belonging to the royal family. As befits social animals, they live in a large, well-maintained group and enjoy extensive time each afternoon and night in the jungle. Since the center relies on income from tourism, nearly all of the elephants must engage in some relatively light work in the morning, such as demonstrations of logging techniques or giving rides through the forest.

The idea for the orchestra was planted in 1999 in New York, when I met Richard Lair, an American who has lived in Southeast Asia for about fifty years. Richard, known as "Professor Elephant" in the region, is the author of *Gone Astray*, the primary reference on the Asian domesticated elephant and the standard elephant veterinary manual. When I met him, he hadn't visited the United States in over nineteen years.

Richard had recently inaugurated a painting project for elephants with the aid of the Russian émigré New York artists Vitaly Komar and Alex Melamid, who had painted with Renee, an elephant in the Toledo, Ohio, zoo in 1995, with help from Don RedFox, the zoo's elephant manager. The artists proposed to establish "Elephant Art Academies" for out-of-work elephants to raise money for the animals and their mahouts (trainers and caretakers of domesticated Asian elephants). At the TECC, the team and the elephant's mahouts quickly taught the elephants to paint abstract designs. The paintings raise significant support for the center and increase awareness of conservation efforts, particularly for the most important audience, the Thai public. Numerous other centers have followed suit, and elephant paintings are at present widely sold by public and private centers.

Komar and Melamid's Thai-speaking guide Linzy Emery arranged for Richard to stay in my apartment. We both enjoy listening to music, and one evening after hours of listening to Junior Kimbrough, Aretha, and Maurice Ravel—and an unaccustomed dram from Scotland's fair shores—we began to wonder whether elephants could learn to play music. Richard told me that the elephant's mahouts know that elephants like to listen to music; they often sing to or play an instrument for the elephants as they walk together through the jungle, and the elephants are calmed. Elephants, moreover, are social animals and might enjoy an activity like playing music together.

While there was good reason to suppose that this idea could work, there were also many questions to consider. Do elephants have any comprehension of music? We read that they did. Rickye and Henry Heffner at the University of Kansas used a simple food reward experiment to elicit an Indian elephant's ability to distinguish simple two-note melodies. They found that elephants could distinguish microtonal pitch gradations smaller than the half-steps on the piano. A study by Karen McComb from the University of Sussex even showed that wild elephants in Kenya could distinguish the ethnicity (the Maasai tribe, who herd cattle and chase elephants away, in contrast to the Kamba tribe, farmers who do not), age, and gender of humans from the sound of their voice.

What sort of instruments should we make for elephants? The instruments should be suitable to the elephant's anatomy, which means they should be large and operable by a trunk. They should also sound Thai, because the regular daily audience is from Thailand (although as of this writing, for economic reasons most of the audience is from China), the mahouts would enjoy the music more, and the elephants have heard Thai music all their lives. Thai tourists who stumbled on the practice and recording sessions told me that the elephants sounded as if they were performing a style of music that can be heard in the Thai Buddhist temples, which I took as a good sign.

We constructed and adapted many instruments, some of which never worked, often because they weren't easy for the elephants to play.

The mahouts told us the elephants especially enjoyed playing the large marimba-like Thai *renaats*. To make the large metal instruments, including the elephant version of the renaat, I mostly worked with Sakhorn (northern Thais generally use only a single name), a talented metalworker in Lampang, about thirty kilometers from the center. I chose a pentatonic musical scale that would suggest traditional northern Thai music (e.g., D E G A C), and in some cases added two American blues notes (e.g., E♭ and B♭).

MATH BOX 11.2
Designing Elephant Xylophones

The standard Thai renaat is a marimba that produces its sounds via vibrating wooden slats. The standard instrument is too delicate to withstand elephants or being left outside during monsoon season. Sakhorn and I constructed the elephant renaat with stainless-steel tubes used to protect fuel and heat lines.

The relationship of the length of a hollow tube and the frequency of sound is different than for a string, in which dividing the length by two produces the next higher octave (see chapter 2). For a tube, the pitch changes with the square root of the length. For example, if the fundamental f_1 is a two-meter tube that sounds middle C (C4) and you want a tube the next octave higher (C5), the length would be

$$2 \text{ m} \times \sqrt{(1/2)} = 2 \text{ m} \times 0.707 = 1.414 \text{ m}.$$

The most common scale I heard in northern (Lanna) Thailand near Lampang is a pentatonic scale consisting of a fundamental, minor third, fourth, fifth, seventh, and octave. We adopted this scale (e.g., E G A B D) for our renaats. I chose a just intonation (see chapter 2) and applied the square root relationship to make relative tube lengths (see table 11.1).

We then needed to attach the tubes to the body of the instrument, and it would be best to do that at the motion nodes (see chapter 3) so that the rest of the tube can vibrate to its maximum extent and produce a pleasing ring.

In a string, the nodes are at the nut and bridge (at the two ends), or where you depress the string with a finger, and the largest vibration is in the middle. In a bar or tube, the greatest motion is not only in the middle but also near the two edges, so the holes should be drilled where the nodes are between these three points. This turns out to occur approximately 2/9 (22.2 percent) of the distance from the ends. For hanging elephant tubular bells, one hole can be placed at this distance and threaded with rope to suspend them. For the renaats, we drilled a hole at each end and tied them with rope to a frame at those points.

TABLE 11.1

Scale name	Frequency ratio	Tube length
Fundamental	1/1	1
Minor third	6/5	0.913
Perfect fourth	4/3	0.866
Perfect fifth	3/2	0.816
Minor seventh	9/5	0.745
Octave	2/1	0.707

Most remarkably, the elephants learn on their own to beat the tubes of renaats at the antinode, that is, the "sweet spot" where the tone sounds clearest, and they avoid hitting near the nodes.

Richard, the mahouts, and I constructed giant tuned slit drums. We made a gong from a circular saw blade confiscated from an illegal logging operation. The harmonica maker Lee Oskar donated harmonicas, and we adopted a reed mouth organ (*kaen*) played in the neighboring region, Issan, in northeastern Thailand: I cut the lengths of the reeds of the kaen to match the orchestra's tuning. Together with Neepagong, the director of the woodworkers at the Conservation Center, we constructed giant slit drums and an enormous string instrument that sounds like an electric bass, the *diddley bow*, which inspired Bo Diddley's stage name. Our instrument was, in turn, named after him.

The instrument builder Ken Butler and I chose harmonicas and premade instruments that would mesh with the specially built instruments' scale, and the artist Don Ritter designed a synthesizer. The elephants took easily to the harmonica, which sparked the first elephant music fad: one morning I arrived to hear the sound of harmonicas from all around, from the hills and the river. The elephants walking through the forest were playing harmonicas, which they hold easily in the tip of their trunk. They sometimes blow them into their own ears.

The gong and thunder sheet initially scared some elephants, but they soon got used to them. The kaens worked well for sound production, but the elephants couldn't hold them and needed to use the mahouts as instrument stands. The elephants didn't seem interested in the bells, theremin, or synthesizer keyboard but would play when asked. They disliked playing wind instruments with a large mouthpiece (or, better, trunk piece). A mahout told me they were afraid that a snake might jump through the wind holes and into their trunks!

On our first trip, Mei Kot, then an eight-thousand-pound seventeen-year-old girl, was first frightened by the gong, but around the third afternoon of her performance, we couldn't get her to stop playing it. Her mahout would take the mallet out of her trunk, but she would pick it up and continue playing. This can be heard in the delayed ending of some of the pieces.

Rory Young, a recording engineer, constructed a studio in a clearing in the jungle, and a team of recording engineers drove up from Bangkok. During the course of six trips over later years, we have tried out about forty instruments; another instrument we found that worked well were angalungs, tuned rattles that northern Thai children learn to play in grade school.

One of the successes in incorporating the local culture was to record schoolchildren from a nearby village performing a nursery rhyme about elephants, "Chang Chang Chang," and then arranging it into a big orchestrated piece on our second CD. In each public show at the center, that piece is blasted over the PA system as the elephants pass through the viewing stands built in the teak forest.

It was gratifying to see the mahouts become more and more interested in the Thai Elephant Orchestra, teaching their elephants, inventing new instruments, and expanding the size of the band from the original six to perhaps eighteen players in total. Indeed, the group—by weight—is the largest orchestra in the world, at several times the combined weight of the Berlin Philharmonic.

The orchestra indeed raised public awareness for the center, particularly where it counts most, in Thailand. Sometimes the orchestra plays with human musicians, leading to an infamous concert where I arranged the first movement of Beethoven's Sixth Symphony, the *Pastoral Symphony*, for the elephants and a sixty-piece school marching band from the Galyani middle school in Lampang, conducted by the band's teacher, Rakhorn. We were proud that a BBC simulcast of the concert was picked up by the American TV show *Jon Stewart's Daily Show* as an episode's closing "moment of zen."

The local Thai newspaper article wrote:

> The Thai Elephant Orchestra, conducted by Richard Lair, had a command performance for HM Queen Sirikit of Thailand. The concert was a collaboration with the 60-strong concert band of Galayani School. The warmly received performance started with the elephants on their own; subsequently, the children's orchestra came in, selected elephants jammed along with it to then segue into elephants only. Most of the music was Thai but there were two western songs, "I Did It My Way" (a favorite of Her Majesty) and "Oh Danny Boy."

I don't think it's interesting to teach elephants to play prewritten human melodies. It's much more interesting to hear how they choose to play. After teaching the elephants to play the instruments and giving some indication of how the instrument should be played for that piece, Richard or I would cue the elephant and mahout to start and stop. The mahout would encourage his animal by moving his arms in a mime of the elephant's trunk.

Except for "Chang Chang Chang," a thirty-two-note melody that the mahouts on their own taught the elephants to perform on the angalung, the notes and rhythms of the pieces are chosen completely by the elephants (figure 11.4). One surprise is that they play variously in duple meter (straight

FIGURE 11.4 The Thai Elephant Orchestra

Members of the Thai Elephant Orchestra performing on renaats and, in the standing percussion section, on a Thai temple gong, a gong made from an illegal confiscated logging saw blade, and a set of tuned tubular bells.

Source: Photos courtesy Millie Young, Mahidol University International College. Used with permission.

eighth notes), triple meter (alternating quarter and eighth notes), and a dotted rhythm (dotted eighth and sixteenth). Sometimes they found motifs for a particular piece and repeated them. I cannot say why they made these choices.

On returning to the United States, many people asked me: is this music? I propose an answer based on the Turing test, which was designed to determine if a computer possesses intelligence. Play the recording for people who don't know the identity of the performers and ask them if it's music. They may love it or beg you to stop, but I think they will say that "of course it's music." I tried this once with a music critic from the *New York Times*, who eventually guessed, "It's an Asian group." He was initially upset when I told him who the performers were, but the next day he contacted me and asked to write about the orchestra.

I'm also confident that the elephants understand the connection between many of their actions and the sounds they produce. They don't operate the instruments randomly but aim for where the sound is best. One can watch this process occur over time on the renaats. The elephants can easily keep a fairly steady beat on numerous instruments, and in the case of Luk Kop ("Tadpole") can alternate between several drums. Long before, Richard had trained Luk Kop to be the elephant in a Disney movie, *Operation Dumbo Drop*. He can be a very sweet elephant who loves to be fed candy and have his tongue petted, but as an enormous adult he is sometimes very dangerous—but only to humans, never to other elephants, and never to his chief mahout. He has, however, chased other people through the forest. So Luk Kop has been retired from the orchestra and all other forms of activity where he might encounter tourists. Still, he was our most avid and talented drummer.

I suspect that at least some of the elephants enjoy playing instruments that are well tuned and have a pleasant resonance. As mentioned, elephants learn on their own to hit renaat bars on their sweet spot, where a more resonant tone is produced, rather than near the nodes, which produces a sharp percussive clank. On a renaat, I planted a dissonant note to see what would happen, and the filmmaker Kurt Ossenfort recorded Pratidah playing on videotape. For the first several minutes she avoided the note but later would not stop playing it. Perhaps like a punk rocker or early-twentieth-century composer, Pratidah had discovered a pleasant dissonance. At any rate, she certainly outsmarted my test design.

On one trip, I was accompanied by Aniruddh Patel, a neuroscientist at Tufts who specializes in animal musicality. Aniruddh felt that while the elephants can actually keep a steadier beat than most humans on the drums, they may not be playing in synchrony at all. I am not convinced, however, as it appears that the elephants often play in offbeat and triplet rhythms with each other, which would be counted as "out of sync." As with Pratidah's dissonance, we may simply not have designed the right test or analysis for such smart animals.

I have seen a few instances when an elephant will walk over to an instrument spontaneously and play it for a few seconds, and Richard and the mahouts tell me this is fairly common. I suspect that this is when the elephants are bored. I have not seen two elephants spontaneously perform together. The urging by the mahouts can range from a single word to ongoing ear tugs for every movement. Some elephants, primarily Poong and Pratidah, walk over to a renaat and play a solo with no urging during the appropriate time during the daily logging show and with no mahout near them and decide themselves when to begin and end their performance.

Pratidah was originally the outstanding renaat player, coming up with beautiful phrases and melodies. Sometimes she would refuse to stop playing her renaat. As with Mei Kot, it is very hard to get her to stop if she doesn't want to!

An outstanding example of one of such solo, played without any instruction during the performance (albeit with a single edit in the recording studio), is from Poong on the Thai Elephant Orchestra's second CD, *Elephonic Rhapsodies*. The sole instruction by humans was to give Poong a stick to initiate the solo, and Poong himself decided when to end the music. When he finished, he simply dropped the stick and walked away. It was on a cadence, where it sounds to us humans like the song should end. The solo has a repeated theme and rhythm, but the extent to which this is random, ergonomic, or intended is guesswork. There is still the remarkable situation where a nonhuman animal is creating his own beautiful music intentionally on an instrument, whatever the underlying mental processes and our extent of understanding.

This brings us to parting troubling thoughts. Some say that it is wrong that nonhuman animals be trained to perform humanlike activities. One should certainly question the notion of teaching elephants to perform human tasks, whether in warfare, logging, riding, babysitting (!), or art and sport, all of them obviously not a part of their wild behavior. I agree with this belief at least for those species that we have not selectively bred—I think domestic horses can love being ridden once they are trained, for example—but only if healthy wild behavior is a genuine alternative. The Asian elephant has been domesticated and trained for humanlike activities for thousands of years, and light work performing for tourists in activities that are enjoyable, like playing in the orchestra, is preferable to using the elephants as war machines, trucks, and loggers.

I believe, as does Richard Lair, that the optimal situation would be to have large wild spaces for the elephants to live in herds, with a means for us to view them without interfering with them except for situations such as medical intervention—perhaps similar to how Kenya provides well-run game parks such as the Maasai Mara. But we are certain that this will not occur to any reasonable extent in our lifetimes, and steps must be taken now to help the long-term survival of the species as well as to support out-of-work elephants and their mahouts, who cannot otherwise support themselves. I think it likely that the long-term survival of the center and other such camps for elephants may become extremely important for maintaining genetic diversity to avoid diseases that would possibly lead to extinction as both domestic and wild populations dwindle. For far more detail on the sad facts behind this situation, I recommend Richard Lair's book *Gone Astray* (published by the Food and Agriculture Organization, a division of the United Nations). And to learn about and enjoy the wonderful elephants and their millennia-old relationship with our species, consider a visit to the center or any of the other well-run conservation centers in Southeast Asia.

A further troubling concern raised by this type of animal training is the temptation to make spurious claims for humanlike behavior. For instance, after Komar and Melamid and Richard Lair introduced elephant painting, some mahouts trained elephants to make figurative paintings, such as trees and flowers, with the implication by some that the elephants know what they are painting. The tourists usually don't realize that this is a "trick," in that the mahout is surreptitiously telling the elephant how to move the brush, often by pressure on a tusk: I have heard but not yet observed that some animals have been trained to do this even without the mahout touching them. The elephant has no notion that it is painting a representative figure. We should guard against these kinds of misunderstandings as they will lead to people suspecting and dismissing the amazing things that elephants are truly capable of.

Whatever fantastic abilities humans presume that elephants may have will be less amazing than the genuine abilities we learn from careful observation. Considering the complexities of how music is heard and created, I think that a great deal of experimentation would be required to discern how elephants experience the music.

Other Musical Species and Us

With the exception of the Cretaceous-Paleogene extinction 66 million years ago, apparently caused by an asteroid that struck Mexico, the current period of extinction of other species, known as the Holocene or Anthropocene extinction, is being caused by our species. Recently there have been population increases of several whale species, for which appreciation of their singing has played a role, but there are still devastating ongoing losses of whale species as of this writing, particularly for the right whale.

The whales, the most exceptional and inspiriting singers on earth; elephants, who rank among the advanced vocal communicators and improvise entire orchestral works; the chimpanzees and bonobos, with skills and cultures we are only starting to comprehend; and many other exceptional and inspiring species are now on a rapid path to extinction and will never be replaced in the entire future of the universe.

If other species produce music the way that we do, does that mean that we must guarantee their future survival rather than destroy them through our thoughtlessness and greed? Why, yes it does.

Art for all species!

—New York City, August 2020

Listening #11

There are few commercial recordings of animal sounds, and most are presently on individual or foundation websites.

Bernie Krause has recorded "soundscapes" in environments throughout the world and offers priceless advice on how to do this yourself.

The monumental website for animal sound is the Macauley Library Archive at the Cornell Lab of Ornithology. It is said to have over seven thousand hours of recordings from over nine thousand species.

The lyrebird should be listened to and viewed, or you will not believe what they can do. The David Attenborough films are stunning. The birds that produce the mechanical noises from humans are in captivity, but the songs derived in part from human flute music are sung by wild birds.

Of all written musical compositions influenced by birdsong, a large body of work by Olivier Messiaen truly stands out. He would take walks into the woods and transcribe the songs and calls as best he could and write them out for the musical instruments available to him. Some pieces are "soundscapes" with birds. Try "Catalogue d'oiseaux," as recorded on the piano by his wife, Yvonne Loriod.

Songs of Insects (http://www.songsofinsects.com) has a great collection of North American cicada, katydid, and cricket songs.

Some recording engineers have slowed the songs of crickets to be more appealing to our species; listen to Thomas Walker's work, on the website of the Department of Entomology at the University of Florida.

My webpage on Animal Music (http://davesoldier.com/animal.html) has recordings of zebra finches triggering musical samples and also hosts recordings by Jay Hirsch of the University of Virginia of fruit fly courtship songs and fruit flies exposed to cocaine aerated in their jar.

Mirjam Knoernschild from the Museum of Natural History in Berlin has recorded an excellent collection of singing and calls by the greater sac-winged bat, which chirps like a songbird in addition to using its voice for echolocation.

The deservedly quintessential animal-vocal hit album is *Songs of the Humpback Whale*, featuring the exemplary recordings by Frank Watlington, Scott McVay, and Roger and Katie Payne.

Recordings of several whale species can be heard from the Monterey Bay Aquarium Research Institute. The blue whale song sped up fivefold is quite gorgeous for us.

The University of Rhode Island's website Discovery of Sound in the Sea, or "DOSITS" (https://dosits.org/), offers numerous sounds of the ocean, including shellfish, fish, and marine mammals such as whales—check out the sea urchin and mantis shrimp as well as the bearded, ribbon, and cra-beater seals.

Bowhead whale songs recorded by the Norwegian Polar Institute can be heard in a 2013 recording in the Fram Strait between Greenland and Norway, available from the *Washington Post*. Among their many voices is a dead-on cat meow.

Listen to the recordings by Sam Ridgway from the National Marine Mammal Foundation of the beluga whale Noc imitating the sounds and cadence of human speech, something he did not do with other belugas.

The Cornell website also has a page on elephants, comparing their low-frequency vocalizations with sine waves at 10, 20, and 30 Hz to explain their long-range communication. Still, most of the sounds made by elephants are in our audible range, and a large vocabulary of Asian elephant vocalizations can be heard on recordings by Richard Lair and myself as *Elephant Field Recordings*.

The Thai Elephant Orchestra recorded three commercial albums. A good solo piece is "Phuong's Solo" on the renaat, and a nice piece with five elephants is "Thung Kwian Sunrise," which is the name of the village where the local mahouts live between Chang Mai and Lampang. I arranged these elephant compositions for human musicians, the first for solo piano, as performed by Steven Beck and Jai Jeffryes, and the second for a full orchestra, as performed by the Composer's Concordance Orchestra.

There are several recordings of the Thai Elephant Orchestra playing along with human musicians, including the local high school brass orchestra on Beethoven's *Pastoral Symphony*; improvised pieces with the cellist Jamie Seiber; the elephants playing along with traditional Thai musicians, including a lovely jam between the mahout string band and the orchestra on "Floating Down the Pin River"; and yours truly playing violin with the elephants on "Little Elephant Saddle."

The simplest and most elegant of the human with elephant musical pieces in my opinion is "Invocation," in which Luk Kop's mahout, Boonyang Boonthiam, who is also a priest of the Lanna Thai religion, sings a traditional elephant prayer with the full orchestra of fourteen elephants.

Acknowledgments

For comments and advice as this book was written, I extend many thanks
to my colleagues:

Nigel Bamford (Yale University), for discussion of neurotransmitter
receptor pathways
Anthony Barnett, for commentary
Gyorgy Buzsaki (New York University), for discussion of the rhythms of
brain activity
Fausto Cattaneo (University of Chicago) for commentary
Marina Combatti (Columbia University), for discussion of ancient
lyric meters
Pedro Cortes, for an explanation of flamenco rhythms
Robert Dick, for discussion of flute and clarinet harmonics
Andrew Farnsworth (Cornell University), for discussion of bird night
flight calls
Brad Garton (Columbia University), for his critique on the entire volume
Wulf Hein (Archaeotechnik, Germany), for insights on ancient
bird-bone flutes
Jay Hirsh (University of Virginia), for discussion of *Drosophila* song
Ben Holtzman (Columbia University), for discussion of sound
wave propagation

Nima Mesgarani (Columbia University), for discussion of cortical processing of speech

Mark Moffet (National Museum of Natural History, Smithsonian), for discussion of insect sound

Manuela Nowotny (University of Jena, Germany), for discussion of insect hearing

Michael Rosen (Columbia University), for discussion of cardiac action potentials and his comments on the entire book

David Rothenberg (New Jersey Institute of Technology), for critique on bird and whale song

Giancarlo Ruocco (Italian Institute of Technology, Rome), for his thorough critique on the entire volume, especially on physics

Susan Savage-Rumbaugh (University of Iowa), for deep discussions on animal music and communication

Ofer Tchernichovski (Hunter College, CUNY), for insights on birdsong

Mark Wightman (University of North Carolina), for comments on the bathtub Ohm's law analogy and general advice throughout the book

Roy Wise (National Institutes of Health), for insights on the history and significance of the dopamine system in behavior

Stanislas Zakharenko (St. Jude Children Research Hospital, Memphis), for insights on hearing voices

I thank the students in our Music, Math, and Mind class who proofed the text, particularly Shashaank N.

I am specially indebted to my friends, including the late Gordon Shaw (University of California, Irvine), for attempts to study music-making pygmy chimpanzees; my collaborator, the conservationist Richard Lair and his colleagues at the Thai Elephant Conservation Center and the Forest Industry Organization of Thailand, for our work on the Thai Elephant Orchestra; and Sue Savage-Rumbaugh, for our ongoing discussions of how to pursue the understanding of music making and perception by other animals.

Thanks to the artists Lisa Haney and Jai Jeffryes, both of whom are outstanding musicians and great friends, for their marvelous original artwork.

Special thanks to Francesca Bartolini, for ongoing inspiration and suggestions.

Permission for extended quotations from James E. Olds is from James L. Olds (George Mason University) and from Mark Moffet, with particular thanks to Susan Savage-Rumbaugh, who prepared her quotation for this book.

Appendix 1

• • • •

Musical Pitch to Frequency Table

In table A.1, the pitch names for a twelve-tone equal temperament scale for conventional A440 tuning are at left, and the octaves are displayed at top. The corresponding frequencies, in Hz, appear in the cells of the table.

Middle C (named from the four-octave organ keyboard) is C4, with A0 being the lowest and C8 the highest note of a conventional eighty-eight-key piano keyboard.

The fundamental frequency f_1 of C0 is below the conventional eighty-eight-key piano keyboard and below our range to hear as a musical note; we perceive only the harmonics. The lowest pitch on a conventional string or electric bass is E1 (41 Hz) or, with a fifth low B string, 31 Hz. The highest note f_1 on a piccolo is C8 and is pretty disturbing to listen to, but even higher frequencies, up to about E♭10, contribute perceptibly to speech and the sound of human music.

The approach that can be used to calculate this table and analogous ones for other approaches to tuning, including dividing the scale into different numbers of equal parts or by the natural overtone series, is discussed in chapter 2.

TABLE A.1

Octave	0	1	2	3	4	5	6	7	8	9
C	16	33	65	131	262	523	1,047	2,093	4,186	8,372
C♯	17	35	69	139	277	554	1,109	2,217	4,435	8,870
D	18	37	73	147	294	587	1,175	2,349	4,699	9,397
E♭	19	39	78	156	311	622	1,245	2,489	4,978	9,956
E	21	41	82	165	330	659	1,319	2,637	5,274	10,548
F	22	44	87	175	349	698	1,397	2,794	5,588	11,175
F♯	23	46	92	185	370	740	1,480	2,960	5,920	11,840
G	24	49	98	196	392	784	1,568	3,136	6,272	12,544
A♭	26	52	104	208	415	831	1,661	3,322	6,645	13,290
A	28	55	110	220	440	880	1,760	3,520	7,040	14,080
B♭	29	58	117	233	466	932	1,865	3,729	7,459	14,917
B	31	62	123	247	494	988	1,976	3,951	7,902	15,804

Appendix 2

• • • •

Further Reading

The Egyptologist Rita Lucarelli, my collaborator on the opera *The Eighth Hour of Amduat*, took me on a walking tour through the Egyptian wing of the Metropolitan Museum of Art in New York, where she read the hieroglyphs from five-thousand-year-old papyri.

I, however, cannot read computer drives from eight years ago. Websites simply disappear, sometimes even over the course of the year I spent writing this volume.

Over the past decade, multiple libraries at my university have closed, the books destroyed. I hope as you pursue this study that you can find whatever the media in whichever resources survive.

Because of their transience, I limit mention of websites as much as possible, although some, like the Cornell McCauley lab's birdsong website, are irreplaceable, and I will take the chance.

For those interested in pursuing scientific topics, I make efforts to mention the names of the scientists who published specific findings. The most convenient way to find the primary articles is to enter their names into PubMed, an enormous website maintained by the National Institutes of Health. While some journals require exorbitant fees to read their old literature, many are open access. Often there are review articles that cover the field well.

The definitive book on these topics is Herman von Helmholtz's *On the Sensations of Tone*, first published in 1863. I have returned to it for about thirty-five years and always find an insight I'd previously missed. I give it my highest possible recommendation.

Fantalezioni di musimatefisi, by the musician and University of Rome physicist Paolo Camiz, is a series of Platonic dialogues on music held between an invented Pythagorean philosopher, Melanzio, and his students Alfo, Basilio, and Cleo. Using their discussions, he explains some of the foundations of modern physics. At present, the book is only available in Italian.

Books that concentrate more on cognitive psychology's analysis of music include J. G. Roederer's *The Physics and Psychophysics of Music: An Introduction* and Diana Deutsch's *The Psychology of Music*.

Bart Hopkin has written many books on homemade instrument building.

Nicolas Collins has a cool book, *Handmade Electronic Music*, on analog circuits that can be used to make sound.

Harry Partch's *Genesis of a Music* is idiosyncratic, inspirational, and brilliant, with an introduction by the intrepid Otto Luening.

On microtones and the blues, explore Alan Lomax's *The Land Where the Blues Began*, and to read how experts disagree on any topic, look at Robert Palmer's *Deep Blues*. When American culture disappears from the face of the earth, it could be reconstructed from these two books.

For the study of syncopated drum patterns, I think the exemplary studies are the collections of West African transcriptions and analysis by Reverend A. M Jones.

For ideas on rhythmic periodicities, you might explore Jozef Schillinger's *System of Musical Composition*.

For the history of how West African and European music crossed and mutated in the New World, read Ned Sublette's *Cuba and Its Music*.

The Interspecies Conversations project is, among other things, a resource for articles on animal communication, including music.

For birdsong, see *Nature's Music: The Science of Birdsong*, by Peter R. Marler and Hans Slabbekoorn, and Hollis Taylor's *Is Birdsong Music?*

For a thorough understanding of conservation and cultural issues for the Asian elephant, read *Gone Astray: The Care and Management of the Asian Elephant in Domesticity*, by Richard Lair, published by the Food and Agricultural Organization of the United Nations.

Bibliography

Introduction

Bach, Johann Sebastian. *Precepts and Principles for Playing the Through Bass or Accompanying in Four Parts*. Leipzig. 1738.

Neidt, Fredrich Erhard. *Musical Guide*. Hamburg: Johann Mattheson, 1700–1721.

Chapter 1

Doppler, Christian. *On the Coloured Light of the Binary Stars and Some Other Stars of the Heavens—Attempt at a General Theory Including Bradley's Theorem as an Integral Part*. Prague Borrosch and Andre, 1842.

Einstein, Albert. "On the Electrodynamics of Moving Bodies." *Annalen der Physik* 17 (1905): 891–921.

Galilei, Galileo. *Discourses and Mathematical Demonstrations Relating to Two New Sciences*. Leiden: House of Elzevir, 1638.

Hubble, Edwin. "A Relation Between Distance and Radial Velocity Among Extra-Galactic Nebulae." *Proceedings of the National Academy of Sciences of the USA* 15 (1929):168–173.

Mersenne, Marin. *Harmonie universelle, contenant la theorie et al pratique de la musique*. Paris: Chez Sebastien Cramoisy, 1638.

Chapter 2

Chandogya Upanishad. 800–600 BCE.

Fabian, A.C., J. S. Sanders, S. W. Allen, et al. "A Deep Chandra Observation of the Perseus Cluster: Shocks and Ripples." *Monthly Notices of the Royal Astronomical Society* 344 (2003): L43–L47.

Farabi, Kitab Al-. *Al-Musiqa al-Kabir* (*The Book of Music*). Circa 950.

Kepler, Johannes. *Harmonices Mundi*. Linz: Godofredi Tampachii, 1619.

Mishaqa, Mikhail. *Essay on the Art of Music for the Emir Shihā*. Circa 1840.

Munzel, Susanne C., Friedrich Seeberger, and Wulf Hein. "The Geissenkloesterle Flute—Discovery, Experiments, Reconstruction." *Studies of Musikarchaologie* 3 (2015).

Partch, Harry. *Genesis of a Music*. Madison: University of Wisconsin Press, 1949.

Von Helmholtz, Hermann. *On the Sensations of Tone as a Physiological Basis for the Theory of Music*. Braunschweig: Friedrich Vieweg and Son, 1863.

Chapter 3

Cherry, E. Colin. "Some Experiments on the Recognition of Speech, with One and with Two Ears." *Journal of the Acoustical Society of America* 25 (1953): 975–79.

Chladni, Ernst. *Entdeckungen über die Theorie des Klanges* [Discoveries in the theory of sound]. Leipzig: Weidmanns Erben and Neid, 1787.

Einstein, Albert. "The Fundamentals of Theoretical Physics." *Science* 91 (1940): 487–92.

Strutt, John William, Baron Rayleigh. *The Theory of Sound*. New York: Macmillan, 1877.

Chapter 4

Anonymous (circa 1426 BCE). *Amduat*. Tomb of Thutmose III, Valley of the Kings.

Brown, Robert. "A Brief Account of Microscopical Observations Made on the Particles Contained in the Pollen of Plants." *London and Edinburgh Philosophical Magazine and Journal of Science* 4 (1828): 161–73.

Einstein, Albert. "Investigations on the Theory of Brownian Movement." *Annalen der Physik* 322 (1905): 549–60.

Fourier, Joseph. *The Analytical Theory of Heat*. Paris: Firmin Didot Père et Fils, 1822.

Galilei, Galileo. *Discourses and Mathematical Demonstrations Relating to Two New Sciences*. Leiden: House of Elzevir, 1638.

Newton, Isaac. *First Book of Opticks*. London: Sam Smith and Benjamin Walford, Printers in the Royal Society at the Prince's Arms in St. Paul's Churchyard, 1704.

Russolo, Luigi. *The Art of Noises*. Milan: Edition Furturiste di Poesia, 1916.

Chapter 5

Ibn al-Khatib of Granada (1313–1374). *Diwan.*

Jones, A. M. *Studies in African Music.* Oxford: Oxford University Press, 1959.

Komaros, Vitsentzos. *Erotokritos* (circa early 1600s). Published in a 1713 edition in Venice.

Mandelbrot, Benoit. *The Fractal Geometry of Nature.* W. H. Freeman, 1982.

Sappho (630–570 BCE). *Collected Poems.*

Schillinger, Joseph. *Schillinger System of Musical Composition.* New York: Carl Fischer, 1946.

Slutsky, Allan ("Dr. Licks"), and Chuck Silverman. *The Funkmasters: The Great James Brown Rhythm Sections, 1960–1973.* New York: Alfred Music, 1997.

Tennyson, Alfred. "The Charge of the Light Brigade." *Examiner,* December 9, 1854.

Wordsworth, William. "My Heart Leaps Up." In *Poems, in Two Volumes.* London: Longman, Hurst, Ress and Orme, 1807.

Chapter 6

Berger, Hans. "Über das elektroenkephalogramm des menschen." *Archiv für Psychiatrie und Nervenkrankheiten* 87 (1929): 527–70.

Bernstein, Julius. "Untersuchungen zur Thermodynamik der bioelektrischen Strome." *Pflügers Archiv* 92 (1902): 21–562.

Cavendish, Henry. "An Account of Some Attempts to Imitate the Effects of the Torpedo by Electricity." *Philosophical Transactions* 66 (1776): 196–225.

Changeux, Jean-Pierre. "Golden Anniversary of the Nicotinic Receptor." *Neuron* 107 (2020): 14–16.

Dale, Henry. "On Some Physiological Actions of Ergot." *Journal of Physiology* 34 (1906): 163–206.

Galvani, Luigi. *De viribus electricatatis in motu musculari.* Bononia: Ex Typographia Instituti Scientiarium, 1791.

"Galvanic Miracles." *Morning Post* (London), January 6, 1803.

Goldman, David. "Potential, Impedance, and Rectification in Membranes." *Journal of General Physiology* 27 (1943): 37–60.

Hodgkin, Alan L., and Andrew F. Huxley. "A Quantitative Description of Membrane Current and Its Application to Conduction and Excitation in Nerves." *Journal of Physiology* 117 (1952): 500–544.

Katz, Bernard. 1966. *Nerve Muscle and Synapse.* New York: McGraw-Hill.

Langley, J. N. "On the Reaction of Cells and of Nerve-Endings to Certain Poisons, Chiefly as Regards the Reaction of Striated Muscle to Nicotine and to Curari." *Journal of Physiology* 33 (1905): 374–413.

Loewi, Otto. "Über humorale übertragbarkeit der Herznervenwirkung." *Pflügers Archiv* 189 (1921): 239–42.

Matteuci, Carlo. "Results of Researches on the Electric Function of the Torpedo." *Proceedings of the Royal Society of London* 10 (1860): 576–79.

Nozaradan, S., I. Peretz, and A. Mouraux. "Selective Neuronal Entrainment to the Beat and Meter Embedded in a Musical Rhythm." *Journal of Neuroscience* 32 (2012): 17572–81.

Palay, Sanford, and George Palade. "The Fine Structure of Neurons." *Journal of Biophysics Biochemistry Cytology* 1 (1955): 69–88.

Sames, D., M. Dunn, R. J. Karpowicz, and D. Sulzer. "Visualizing Neurotransmitter Secretion at Individual Synapses." *ACS Chemical Neuroscience* 4 (2013): 648–51.

Shelley, Mary. *Frankenstein; or, The Modern Prometheus*. Introduction to the 1831 ed. London: Henry Colburn and Richard Bentley.

Sherrington, Charles. *The Integrative Action of the Nervous System*. London: Archibald Constable and Co., 1906.

Sulzer, D., and E. N. Pothos. "Regulation of Quantal Size by Presynaptic Mechanisms." *Reviews in the Neurosciences* 11 (2000): 159–212.

Sulzer, Johann Georg. *Recherches sur l'origine des sentimens agréables et désagréables*. Troisième partie: *Des plaisirs des sens*. Histoire de l'Académie Royale des Sciences et des Belles-Lettres de Berlin. Berlin: Haude et Spenner, 1752.

Volta, Alessandro. "Account of Some Discoveries Made by Mr. Galvani, with Experiments and Observations on Them." *Philosophical Transactions* 83 (1793): 10–44.

Winkler, I., G. P. Háden, O. Ladinig, et al. "Newborn Infants Detect the Beat in Music." *Proceedings of the National Academy of Sciences of the USA* 106 (2009): 2468–71.

Chapter 7

Bamford, Nigel, R. Mark Wightman, and David Sulzer. "Dopamine's Effects on Corticostriatal Synapses During Reward-Based Behaviors." *Neuron* 97 (2018): 494–510.

Blood, A. J., and Robert Zatorre. "Intensely Pleasurable Responses to Music Correlate with Activity in Brain Regions Implicated in Reward and Emotion." *Proceedings of the National Academy of Sciences of the USA* 98 (2001): 11818–23.

Carlsson, Aarvid. "3,4-Dihydroxyphenylalanine and 5-Hydroxytryptophan as Reserpine Antagonists." *Nature* 180 (1957): 1200.

Carlsson, Aarvid, and M. Lindqvist. "Effect of Chlorpromazine or Haloperidol on the Formation of 3-Methoxytyramine and Normetanephrine in Mouse Brain." *Acta Pharmacologica* 20 (1963): 140–44.

Day, Jeremy, Mitchell Rotiman, R. Mark Wightman, and Regina M. Carelli. "Associative Learning Mediates Dynamic Shifts in Dopamine Signaling in the Nucleus Accumbens." *Nature Neuroscience* 10 (2007): 1020–28.

Olds, James E. "Commentary: The Discovery of Reward Systems in the Brain." In *Brain Stimulation and Motivation: Research and Commentary*, ed. Elliot S. Valenstein. Northbrook, IL: Scott, Foresman, 1973.

——. "Pleasure Centers in the Brain." *Scientific American* 195 (1956): 105–17.

Oliver, George, and E. A. Schäfer. "The Physiological Effects of Extracts of the Suprarenal Capsules." *Journal of Physiology* 18 (1895): 230–76.

Schultz, Wolfram. "Getting Formal with Dopamine and Reward." *Neuron* 36 (2002): 241–63.

Sulzer, D., S. J. Cragg, and M. E. Rice. "Striatal Dopamine Neurotransmission: Regulation of Release and Uptake." *Basal Ganglia* 6 (2016): 123–48.

Yokel, R. A., and Roy Wise. "Increased Lever Pressing for Amphetamine After Pimozide in Rats: Implications for a Dopamine Theory of Reward." *Science* 187 (1975): 547–49.

Chapter 8

Corti, Alfonso. "Recherches sur l'organe de l'ouïe des mammiferes." *Zeitschrift für wissenschaftliche Zoologie* (1851).

Curie, Jacques, and Pierre Curie. "Contractions and Expansions Produced by Voltages in Hemihedral Crystals with Inclined Faces." *Comptes Rendus* 93 (1881): 1137–40.

Dallos, Peter, Jing Zheng, and Mary Ann Cheatham. "Prestin and the Cochlear Amplifier." *Journal of Physiology* 576 (2006): 37–42.

Hudspeth, A. J. "Integrating the Active Process of Hair Cells with Cochlear Function." *Nature Reviews Neuroscience* 15 (2014): 600–14.

Khalegi, Morteza, Curlong Cosme, Mike Ravica, et al. "Three-Dimensional Vibrometry of the Human Eardrum with Stroboscopic Lensless Digital Holography." *Journal of Biomedical Optics* 20 (2015): 051028.

Olson, Elizabeth S., Hendrikus Dukfhuis, and Charles T. Steele. "Von Békésy and Cochlear Mechanics." *Hearing Research* 293 (2012): 31–43.

Olson, Elizabeth S., and Manuela Nowotny. "Experimental and Theoretical Explorations of Traveling Waves and Tuning in the Bushcricket Ear." *Biophysical Journal* 116 (2019):165–77.

von Békésy, G. "Current Status of Theories of Hearing." *Science* 123 (1956): 779–83.

——. *Experiments in Hearing.* New York: McGraw-Hill, 1960.

——. "Pendulums, Traveling Waves, and the Cochlea: Introduction and Script for a Motion Picture." *Laryngoscope* 68 (1958): 317–27.

——. "Travelling Waves as Frequency Analysers in the Cochlea." *Nature* 225 (1970): 1207–9.

Von Helmholtz, Hermann. 1863. *On the Sensations of Tone as a Physiological Basis for the Theory of Music.* Braunschweig: Friedrich Vieweg and Son.

Chapter 9

Abrams, D. A., T. Chen, P. Odriozola, et al. "Neural Circuits Underlying Mother's Voice Perception Predict Social Communication Abilities in Children." *Proceedings of the National Academy of Sciences of the USA* 113 (2016): 6295–300.

Chen, Liang, Xinxing Wang, Shaoyu Ge, and Qiaojie Xiong. "Medial Geniculate Body and Primary Auditory Cortex Differentially Contribute to Striatal Sound Representations." *Nature Communications* 10 (2019): 418.

Di Liberto, G. M., C. Pelofi, R. Bianco, et al. "Cortical Encoding of Melodic Expectations in Human Temporal Cortex." *Elife 9* (2020): e51784.

Guo, Wei, Amanda R. Clause, Asa Barth-Maron, and Daniel B. Polley. "A Corticothalamic Circuit for Dynamic Switching Between Feature Detection and Discrimination." *Neuron* 95 (2017): 180–94.

Homma, N. Y., M. Happel, F. R. Nodal, et al. "A Role for Auditory Corticothalamic Feedback in the Perception of Complex Sounds." *Journal of Neuroscience* 37 (2017): 6149–61.

Hubel, David H., and Torsten N. Wiesel. "Ferrier Lecture: Functional Architecture of Macaque Monkey Visual Cortex." *Proceedings of the Royal Society B* 198 (1977): 1–59.

Knudsen, Eric. "Instructed Learning in the Auditory Localization Pathway of the Barn Owl." *Nature* 417 (2002): 322–28.

Konishi, Masakazu. "Coding of Auditory Space." *Annual Reviews Neuroscience* 28 (2003): 31–55.

McAlonan, Kerry, Verity J. Brown, and E. M. Bowman. "Thalamic Reticular Nucleus Activation Reflects Attentional Gating During Classical Conditioning." *Journal of Neuroscience* 20 (2000): 8897–901.

Menegas, William, Kortecki Akiti, Ryunosuke Amo, et al. "Dopamine Neurons Projecting to the Posterior Striatum Reinforce Avoidance of Threatening Stimuli." *Nature Neuroscience* 21 (2018): 1421–30.

Mesgarani, N., C. Cheung, K. Jonson, and E. F. Chang. "Phonetic Feature Encoding in Human Superior Temporal Gyrus." *Science* 343 (2014): 1006–10.

Mesgarani, N., S. V. David, J. B. Fritz, and S. A. Shamma. "Mechanisms of Noise Robust Representation of Speech in Primary Auditory Cortex." *Proceedings of the National Academy of Sciences of the USA* 111 (2014): 6792–97.

Nourski, K. V. "Auditory Processing in the Human Cortex: An Intracranial Electrophysiology Perspective." *Laryngoscope Investigative Otolaryngology* 2 (2017): 147–56.

Penfield, Wilder. "Activation of the Record of Human Experience. Summary of the Lister Oration Delivered at the Royal College of Surgeons of England." *Annual Review of the Royal College of Surgeons of England* 29 (1961): 77–84.

——. *The Excitable Cortex in Conscious Man.* Charles C. Thomas, 1958.

——. "The Interpretive Cortex: The Stream of Consciousness in the Human Brain Can Be Electrically Reactivated." *Science* 26 (1959): 1719–25.

Penfield, Wilder, and Phanor Perot. "The Brain's Record of Auditory and Visual Experience." *Brain* 86 (1963): 595–696.

Roth, H., and B. Sommer. "Interview with Brenda Milner, Ph.D., Sc.D." American Academy of Neurology Oral History Project.

Tramo, Mark, Peter Cariani, Bertrand Delgutte, and Louis Braida. "Neurobiological Foundations for the Theory of Harmony in Western Tonal Music." *Annals of the New York Academy of Science* 930 (2001): 92–116.

Warrier, Catherine M., Daniel A. Abrams, Trent G. Nicol, and Nina Kraus. "Inferior Colliculus Contributions to Phase Encoding of Stop Consonants in an Animal Model." *Hearing Research* 282 (2011): 108–18.

Chapter 10

Bolton, Thaddeus. "Rhythm." *American Journal of Psychology* 8 (1894): 145–238.

Brochard, R., D. Abecasis, D. Potter, et al. "The 'Ticktock' of Our Internal Clock: Direct Brain Evidence of Subjective Accents in Isochronous Sequences." *Psychological Science* (2003): 362–66.

Chun, S., J. J. Westmoreland, I. T. Bayazitov, et al. "Specific Disruption of Thalamic Inputs to the Auditory Cortex in Schizophrenia Models." *Science* 344 (2014): 1178–82.

Jaramillo, F., V. S. Markin, and A. J. Hudspeth. "Auditory Illusions and the Single Hair Cell." *Nature* 364 (1993): 527–29.

Knoblauch, August. "On Disorders of the Musical Capacity from Cerebral Disease." *Brain* 13 (1890): 317–40.

McGurk, Harry, and John MacDonald. "Hearing Lips and Seeing Voices." Nature 264 (1976): 746–48.

Peretz, Isabelle. "Neurobiology of Congential Amusia." *Trends in Cognitive Sciences* 20 (2016): 857–67.

Seebeck, August. "Über die Definition des Tones." *Annalen der Physik* 139 (1844): 353–68.

Shaw, Gordon. *Keeping Mozart in Mind.* New York: Academic Press, 2000.

Tartini, Guiseppe. *De' principj dell'armonia musicale contenuta nel diatonico genere: dissertazione.* Padua: Padova Stamperia del Seminario, 1767.

Terhardt, Ernst. "The Concept of Musical Consonance: A Link Between Music and Psychoacoustics." *Music Perception* 1 (1984): 276–95.

Zatorre, Robert J. J. "Neural Specializations for Tonal Processing." *Annals of the New York Academy of Sciences* 930 (2001): 193–210.

Chapter 11

Aldrich, Herbert Lincoln. *Arctic Alaska and Siberia; or, Eight Months with the Arctic Whalers.* New York: Rand, McNally, 1889.

Arcadi, A. C., and W. Wallauer. "They Wallop Like They Gallop: Audiovisual Analysis Reveals the Influence of Gait on Buttress Drumming by Wild Chimpanzees (*Pan troglodytes*)." *International Journal of Primatology* 34 (2013): 194–215.

Briseño-Jaramillo, M., V. Biquand, A. Estrada, and A. Lemasson. "Vocal Repertoire of Free-Ranging Black Howler Monkeys (*Alouatta pigra*): Call Types, Contexts, and Sex-Related Contributions." *American Journal of Primatology* 79 (2017): 10.1002/ajp.22630.

Clarke, Esther, Ulrich H. Reichard, and Klaus Zuberbühler. "The Syntax and Meaning of Wild Gibbon Songs." *PLoS One* 1 (2006): e73.

Damien, J., O. Adam, D. Cazau, et al. "Anatomy and Functional Morphology of the *Mysticete rorqual* Whale Larynx: Phonation Positions of the U-Fold." *Anatomical Record* 302 (2019): 703–17.

Dufour, V., N. Poulin, C. Charlotte, and E. H. Sterck. "Chimpanzee Drumming: A Spontaneous Performance with Characteristics of Human Musical Drumming." *Scientific Reports* 5 (2015): 11320.

Ferrara, C. R., R. C. Vogt, and R. S. Sousa-Lima. "Turtle Vocalizations as the First Evidence of Posthatching Parental Care in Chelonians." *Journal of Comparative Psychology* 127 (2013): 24–32.

Gamba, M., V. Torti, V. Estienne, et al. "The Indris Have Got Rhythm! Timing and Pitch Variation of a Primate Song Examined Between Sexes and Age Classes." *Frontiers in Neuroscience* 10 (2016): 249.

Gans, Carl, and Paul Maderson. "Sound-Producing Mechanisms in Recent Reptiles: Review and Comment." *American Zoologist* 13 (1973): 1195–203.

Gillespie-Lynch, K., P. M. Greenfield, H. Lyn, and S. Savage-Rumbaugh. "Gestural and Symbolic Development Among Apes and Humans: Support for a Multimodal Theory of Language Evolution." *Frontiers in Psychology* 5 (2014): 1228.

Göpfert, M. C., and R. M. Hennig. "Hearing in insects." *Annual Review of Entomology* 61 (2016): 257–76.

Goodall, J. "Tool-Using and Aimed Throwing in a Community of Free-Living Chimpanzees." *Nature* 201 (1964): 1264–66.

Griffin, Donald, and Robert Galambos. "The Sensory Basis of Obstacle Avoidance by Flying Bats." *Journal of Experimental Zoology* 86 (1941): 481–505.

Gupfinger, R., and M. Kaltenbrunner. "The Design of Musical Instruments for Grey Parrots: An Artistic Contribution Toward Auditory Enrichment in the Context of ACI." *Multimodal Technologies and Interactions* 4 (2020).

Hawthorne, Nathaniel. "The Canterbury Pilgrims." In *Snow-Image, and Other Twice-Told Tales.* New York: Ticknor, Reed, and Fields, 1833.

Heffner, R. S., and H. E. Heffner. "Hearing in the Elephant (*Elephas maximus*): Absolute Sensitivity, Frequency Discrimination, and Sound Localization." *Journal of Comparative and Physiological Psychology* 96 (1982): 926–44.

Herbst, C. T., A. S. Stoeger, R. Frey, et al. "How Low Can You Go? Physical Production Mechanism of Elephant Infrasonic Vocalizations." *Science* 337 (2012): 595–99.

Herman, L. M. "The Multiple Functions of Male Song Within the Humpback Whale (*Megaptera novaeangliae*) Mating System: Review, Evaluation, and Synthesis." *Biological Reviews Cambridge Philosophical Society* 92 (2017): 1795–1818.

Hulse, Stewart, J. Humpal, and J. Cynx. "Processing of Rhythmic Sound Structures by Birds." *Annals of the New York Academy of Sciences* 423 (1984): 407–19.

Kelley, D. B., I. H. Ballagh, C. L. Barkan, et al. "Generation, Coordination, and Evolution of Neural Circuits for Vocal Communication." *Journal of Neuroscience* 40 (2020): 22–36.

Ketten, D. R., J. Arruda, S. Cramer, and M. Yamato. "Great Ears: Low-Frequency Sensitivity Correlates in Land and Marine Leviathans." *Advances in Experimental Medicine and Biology* 875 (2016): 529–38.

Krause, Bernie. *The Great Animal Orchestra: Finding the Origins of Music in the World's Wild Places.* Boston: Little, Brown, 2012.

Lair, Richard. *Gone Astray: The Care and Management of the Asian Elephant in Domesticity.* Geneva: Food and Agricultural Organization of the United Nations.

Larom, D., M. Garstang, K. Payne, et al. "The Influence of Surface Atmospheric Conditions on the Range and Area Reached by Animal Vocalizations." *Journal of Experimental Biology* 200 (1997): 421–31.

Lawrence, D. H. "Tortoise Shout." In *Birds, Beasts, and Flowers*. London: Martin Secker, 1923.

Libby, Orin. "The Nocturnal Flight of Migrating Birds." *Auk* 16 (1899): 140–45.

Liu, T., L. Dartevelle, C. Yuan, et al. "Increased Dopamine Level Enhances Male-Male Courtship in *Drosophila*." *Journal of Neuroscience* 28 (2008): 5539–46.

Luther, D., and L. Baptista. "Urban Noise and the Cultural Evolution of Bird Songs." *Proceedings of the Royal Society B* 277 (2010): 469–73.

McComb, K., G. Shannon, K. N. Sayialel, and C. Moss. "Elephants Can Determine Ethnicity, Gender, and Age from Acoustic Cues in Human Voices." *Proceedings of the National Academy of Sciences of the USA* 111 (2014): 5433–38.

Mooney, T. A., R. T. Hanlon, J. Christensen-Dalsgaard, et al. 2010. "Sound Detection by the Longfin Squid (*Loligo pealeii*) Studied with Auditory Evoked Potentials: Sensitivity to Low-Frequency Particle Motion and Not Pressure." *Journal of Experimental Biology* 213 (2014): 3748–59.

Nakano Ryo, Takanashi Takuma, and Surlykke Annemarie. "Moth Hearing and Sound Communication." *Journal of Comparative Physiology A* 20 (2015): 111–12.

Noad, M. J., D. H. Cato, M. M. Bryden, et al. "Cultural Revolution in Whale Songs." *Nature* 408 (2000): 537.

Nottebohm, F., M .E. Nottebohm, and L. Crane. "Developmental and Seasonal Changes in Canary Song and Their Relation to Changes in the Anatomy of Song-Control Nuclei." *Behavioral Neural Biology* 46 (1986): 445–71.

Panova, E. M., and A. V. Agafonov. "A Beluga Whale Socialized with Bottlenose Dolphins Imitates Their Whistles." *Animal Cognition* 20 (2017): 1153–60.

Patel, A., J. Iversen, M. Bregman, and I. Schulz. "Experimental Evidence for Synchronization to a Musical Beat in a Nonhuman Animal." *Current Biology* 19 (2009): 827–30.

Payne, K. B., W. R. Langbauer, and E. M. Thomas. "Infrasonic Calls of the Asian Elephant (*Elephas maxim*)." *Behavioral Ecology and Sociobiology* 18 (1986): 297–301.

Payne, Roger, and Douglass Web. "Orientation by Means of Long-Range Acoustic Signaling in Baleen Whales." *Annals of the New York Academy of Science* 188 (1971): 110–41.

Payne, Roger S., and Scott McVay. "Songs of Humpback Whales." *Science* 173 (1971): 585–97.

Powys, Vicki, Hollis Taylor, and Carol Probets. "A Little Flute Music: Mimicry, Memory, and Narrativity." *Environmental Humanities* 3 (2013): 43–70.

Rivera-Cáceres, K. D., E. Quirós-Guerrero, M. Araya-Salas, et al. "Early Development of Vocal Interaction Rules in a Duetting Songbird." *Royal Society Open Science* 5 (2018): 171791.

Roffman, I., Y. Mizrachi, S. Savage-Rumbaugh, et al. "Pan Survival in the Twenty-First Century: Chimpanzee Cultural Preservation, Rehabilitation, and Emancipation Manifesto." *Human Evolution* 34 (2019): 1–19.

Schachner, A., T. Brady, I. Pepperberg, and M. Hauser. "Spontaneous Motor Entrainment to Music in Multiple Vocal Mimicking Species." *Current Biology* 19 (2009): 831–36.

Shaw, Gordon. *Keeping Mozart in Mind*. New York: Academic Press, 2000.

Stafford, K. M., S. E. Moore, K. L. Laidre, and M. P. Heide-Jørgensen. "Bowhead Whale Springtime Song off West Greenland." *Journal of the Acoustical Society of America* 124 (2008): 3315–23.

Struhsaker, Thomas. "Two Red-Capped Robin-Chats *Cossypha natalensis* Imitate Antiphonal Duet of Black-Faced Rufous Warblers *Bathmocercus rufus*." *Journal of East African Natural History* 106 (2017): 53–56.

Suga, N. "Neural Processing of Auditory Signals in the Time Domain: Delay-Tuned Coincidence Detectors in the Mustached Bat." *Hearing Research* 324 (2015): 19–36.

Suthers, Roderick. "How Birds Sing and Why It Matters." In *Nature's Music: The Science of Birdsong*, ed. P. Marler and N. Slabbekoorn. Amsterdam: Elsevier, 2004.

Tchernichovski, O., P. P. Mitra, T. Lints, and F. Nottebohm. "Dynamics of the Vocal Imitation Process: How a Zebra Finch Learns Its Song." *Science* 29 (2001): 2564–69.

Ten Cate, Carel, and Michelle Spierings. "Rules, Rhythm, and Grouping: Auditory Pattern Perception by Birds." *Animal Behaviour* 151 (2019): 249–57.

Van Doren, B. M., K. G. Horton, A. M. Dokter, et al. "High-Intensity Urban Light Installation Dramatically Alters Nocturnal Bird Migration." *Proceedings of the National Academy of Sciences of the USA* 114 (2017): 11175–80.

Von Muggenthaler, E., P. Reinhart, B. Lympany, and R. B. Craft. "Songlike Vocalizations from the Sumatran Rhinoceros (*Dicerorhinus sumatrensis*)." *Journal of the Acoustical Society of America* 4 (2003).

Author's Selected Compositions
and Discography

Opera

Naked Revolution, an opera in the socialist realist style (1997), with Komar & Mela-
  mid and Maita di Niscemi, libretto.
A Soldier's Story (2002), radio opera with book by Kurt Vonnegut.
The Eighth Hour of Amduat (2015), opera with libretto adapted from *The Book of the
  Amduat*, about 3000 BCE, featuring Marshall Allen of the Sun Ra Arkestra.

Oratorios and Song Cycles

The Apotheosis of John Brown (1990), text adapted from Frederick Douglass.
Smut, a.k.a., Chorea Lascivia (1991), Latin homoerotic medieval lyrics.
Mark Twain's War Prayer (1993), for gospel choir and orchestra or organ.
Ice-9 Ballads (1995), lyrics by Kurt Vonnegut.
Dean Swift's Satyrs for the Very Very Young (2011), twelve pieces for singer, flute,
  viola, and harp, with lyrics from Jonathan Swift featuring Eliza Carthy.

Orchestra

Ultraviolet Railroad (1991), double concerto for violin and cello or piano trio.
Thung Kwian Sunrise (2012), arranged from an improvisation by the Thai Elephant
  Orchestra.
SamulNori (2013).

Bambaataa Variations (2013), concerto grosso for prepared string quartet and orchestra, themes associated with Afrika Bambaataa and the Soulsonic Force.

Stuff Smith's Unfinished Concerto (2017), for violin, piano, and string orchestra; transcription and arrangement for Stuff Smith's working concerto from 1963.

Jaelo (2019), rhapsody for piano and strings or piano solo.

West Memphis, 1949 (2019), big band.

String Quartet

Sequence Girls (1985), string quartet and trap set drums.

Three Delta Blues (1986), string quartet from songs by Robert Johnson, Skip James, and Charlie Patton.

String Quartet #1 The Impossible (1987), string quartet and trap set drums.

String Quartet #2, Bambaataa Variations (1992), for prepared string quartet (bobby pins, hair brushes, combs), with themes associated with Afrika Bambaataa and the Soulsonic Force.

String Quartet #3, The Essential (2011), for string and EEG headband software, after Schoenberg's Second String Quartet, second movement.

Chamber Ensembles

Duo Sonata (1988), for violin and cello.

To Spike Jones in Heaven (1989), for accordion and tape (or CD).

Utah Dances (1990), dance suite for solo saxophone, clarinet, or flute.

Sontag in Sarajevo (1994), for accordion, violin (or clarinet), cello (or bass), and guitar.

The People's Choice Music (with Komar & Melamid, lyrics by Nina Mankin).

"The Most Wanted Song" (1997), for soprano, baritone, electric piano, synthesizer, piano, three guitars, electric bass, trap set, bass drum, violin, cello, soprano and tenor sax.

"The Most Unwanted Song" (1997), for soprano, children's choir, accordion, bagpipe, banjo, flute/piccolo, harmonica, organ, synthesizer, tuba, harp, two bass drums.

East St. Louis, 1968 (1999), for solo viola or string quartet with recording.

Clever Hans (2005), ballet for violin, cello, and harpsichord.

The Complete Victrola Sessions (2010), twelve pieces for violin and piano.

Lewitt Etudes (2015), fifty architectural designs for musicians.

Vienna Over the Hills/Six Violins (1986/2017), six or more violins.

Solo Piano

Five Little Monsters (1985).

Nocturnes (2010).

Variations on Chopin's Minute Waltz (2010), some mathematical variations are playable and some are impossible.

Fractals on the Names of Bach & Haydn (2011).

Letter to Gil Evans (2012).

Girl with a Hat in a Car (2012).

Letter to Skip James (1987/2012).

Phong's Solo (2012), arranged from an improvisation by a member of the Thai Elephant Orchestra.

Jaleo (2019).

Solo Organ

Hockets & Inventions (1990), twelve pieces for solo organ or piano.

Organum (2011), five pieces for solo organ.

Solo Violin

Calo' (2020), six flamenco etudes for solo violin.

Selected Discography

As Leader/Coleader

Calo', for solo violin, performed by Miranda Cuckson (2021).

Jonathan Kane and Dave Soldier, *February Meets the Soldier String Quartet* (2020).

Jaleo, for solo piano, performed by Steven Beck (2020).

Zajal (lyrics from medieval Andalusian poetry in Hebrew and Arabic) (2020).

Naked Revolution (opera with Komar and Melamid) (2018).

The Eighth Hour of Amduat (opera with libretto from ancient Egypt, featuring Marshall Allen, 2016).

Dean Swift's Satyrs for the Very Very Young (lyrics from Jonathan Swift, with Eliza Carthy, 2016).

Jonathan Kane and Dave Soldier, *SoldierKane* (2016).

In Black & White (music for piano, 2015), performed by Steven Beck and the composer.

In Four Color (music for string quartet, 2015), performed by the PubliQuartet and the Soldier String Quartet.

Smash Hits by the Thai Elephant Orchestra (2015).

With Kurt Vonnegut (operas and song cycles, 2015).

The Kropotkins Portents of Love (2015).

Organum (organ music, 2012).

The Complete Victrola Sessions/The Violinist (2011).

Water Music (with the Thai Elephant Orchestra, 2011).

The Kropotkins, *Paradise Square* (2010).

Yol Ku', Inside the Sun: Mayan Mountain Music (children in San Mateo Ixtatan, Guatemala, 2008).

Chamber Music (2007).

Da Hiphop Raskalz (children in East Harlem, 2006).

Soldier Stories (2005), radio play with Kurt Vonnegut.

Elephonic Rhapsodies (with the Thai Elephant Orchestra, 2004).

Soldier String Quartet Inspect for Damaged Gods (2004).

Thai Elephant Orchestra (2001).

Ice-9 Ballads (2001), song cycle with Kurt Vonnegut.

The Kropotkins, *Five Points Crawl* (2000).

The People's Choice Music (with Komar & Melamid, 1997).

The Tangerine Awkestra, *Aliens Took My Mom* (children in Brooklyn, 2000).

Soldier String Quartet with Robert Dick, *Jazz Standards on Mars* (1997).

The Kropotkins (1996).

Soldier String Quartet, *She's Lightning When She Smiles* (1996).

Smut (lyrics of homoerotic poetry in medieval Latin) (1994).

War Prayer (libretto from Mark Twain, 1994).

The Apotheosis of John Brown (libretto from Frederick Douglass, 1993).

Soldier String Quartet, *Sojourner Truth* (1991).

Soldier String Quartet, *Sequence Girls* (1988).

Arranger, Performer, Composer, Conductor, Producer

Vince Bell, *Ojos* (coproducer with Bob Neuwirth & Patrick Derivaz, 2018).

John Cale: *Fragments of a Rainy Season* (1992), *Paris S'Eveille* (1993), *Antartida* (1995), *Walking on Locusts* (1996), *Eat and Kiss* (1997), *Dance Music* (1998), *I Shot Andy Warhol* (1997).

John Clark, *Sonus Innerabalis* (2016), producer.

Nicolas Collins, *A Dark & Stormy Night* (1992).

Pedro Cortes, *Los Viejos No Mueron* (2013), producer.

Sussan Deyhim, *Madman of God* (1999).

Robert Dick, *Third Stone from the Sun* (1993).

Grupo Wara, *Malombo* (1990).

Guided by Voices: *Do the Collapse* (1999), *Hold on Hope* (2000), *Isolation Drills* (2001).

Jason Kao Hwang, *Symphony of Souls* (2011).

Jessie Harris, *While the Music Lasts* (2004), arrangements by Van Dyke Parks.

Jonas Hellborg and Tony Williams, *The Word* (1992).

William Hooker: *Yearn for Certainty* (2010) (trio with Sabir Mateen), *Heart of the Sun* (2013) (trio with Roy Campbell), *Aria* (2016).

Leroy Jenkins, *Themes & Variations on the Blues* (1994).

Sylvain Leroux & L'Ecole Fula: *Flute Les enfants de Tyabla* (2014), *Tyabla* (2019) (executive producer, children in Conakry, Guinea, composing and performing on the fula flute).

Mandeng, *Eletrik* (2004).

Bob Neuwirth and John Cale, *Last Day on Earth* (1994).

Phill Niblock/Soldier String Quartet, *Early Winter* (1994).

Le Nouvelles Polyponies Corses (Corsican Polyphony), *Le Praiduisu* (1999).

The Ordinaires, *The Ordinaires* (1987).

Christina Rosenvinge, *Foreign Land* (2002).

Sequitur, *To Have and to Hold* (2007).

Elliott Sharp/Soldier String Quartet: *Tessalation Row* (1987), *Hammer, Anvil, Stirrup* (1989), *Twistmap* (1991), *Cryptoid Fragments* (1993), *Rheo/Umbra* (1998), *Xeno-Codex* (1996), *String Quartets, 1986–1996* (2003), *Larynx* (1987), *Syndakit* (1999).

Lorette Velvette, *Lost Part of Me* (1998).

Film Scores

Basquiat (dir. Julian Schnabel) (1998), arrangements.

Eat (dir. Andy Warhol, music John Cale) (1995), arrangements.

I Shot Andy Warhol (dir. Mary Herron, music by John Cale) (1997), arrangements.

In Bed with Ulysses (dir. Alan Adelson and Kate Taverna).

Kiss (dir. Andy Warhol, music John Cale) (1995), arrangements.

Mekong Delta (dir. Vanessa Ly) (2003), winner Hong Kong Film Festival.

Serenade (animation by Nadia Roden), winner New York Film Festival (2003).

Special Friends (with Teo Macero) (1988).

Sesame Street (six cartoons for the TV show, dir. Nadia Roden) (1999).

The Violinist (dir. Winsome Brown) (2011).

"Pop" Arrangements and Conducting

David Byrne (Talking Heads), John Cale (Velvet Underground), Guided by Voices, Jonathan Richman (the Modern Lovers), Christina Rosenvinge, Sesame Street, Ric Ocasek (the Cars), Van Dyke Parks (the Beach Boys), Jesse Harris, Alana Amram, Richard Hell (the Voidoids), Sussan Deyhim, Rufus Wainwright, Syd Straw.

Index

Note: photos, figures, tables, and sidebars are indicated by italicized page numbers

Printed in the USA
CPSIA information can be obtained
at www.ICGtesting.com
JSHW050002170524
63278JS00014B/52

9 780231 193795